Multicriteria Analysis for Environmental Decision-Making

THE ANTHEM ENVIRONMENT AND SUSTAINABILITY INITIATIVE (AESI)

The Anthem Environment and Sustainability Initiative (AESI) seeks to push the frontiers of scholarship while simultaneously offering prescriptive and programmatic advice to policymakers and practitioners around the world. The programme publishes research monographs, professional and major reference works, upper-level textbooks and general interest titles.

Professor Lawrence Susskind, as General Editor of AESI, oversees the below book series, each with its own series editor and an editorial board featuring scholars, practitioners and business experts keen to link theory and practice.

Anthem Strategies for Sustainable Development Series
Series Editor: Professor Lawrence Susskind (MIT)

Anthem Climate Change and Policy Series
Series Editor: Dr. Brooke Hemming (US EPA)

Anthem Diplomacy at the Food-Water-Energy Nexus Series
Series Editor: Professor Shafiqul Islam (Tufts University)

Anthem International Environmental Policy Series
Series Editor: Professor Saleem Ali (University of Delaware)

Anthem Big Data and Sustainable Cities Series
Series Editor: Sarah Williams (MIT)

Included within the AESI is the Anthem EnviroExperts Review.

Through this online micro-review site, Anthem Press seeks to build a community of practice involving scientists, policy analysts and activists committed to creating a clearer and deeper understanding of how ecological systems – at every level – operate, and how they have been damaged by unsustainable development. This site publishes short reviews of important books or reports in the environmental field, broadly defined. Visit the site: www.anthemenviroexperts.com.

Multicriteria Analysis for Environmental Decision-Making

Davide Geneletti

ANTHEM PRESS

Anthem Press
An imprint of Wimbledon Publishing Company
www.anthempress.com

This edition first published in UK and USA 2019
by ANTHEM PRESS
75–76 Blackfriars Road, London SE1 8HA, UK
or PO Box 9779, London SW19 7ZG, UK
and
244 Madison Ave #116, New York, NY 10016, USA

British Library Cataloguing-in-Publication Data
A catalogue record for this book is available from the British Library.

Library of Congress Cataloging-in-Publication Data
Names: Geneletti, Davide, author.
Title: Multicriteria Analysis for Environmental
Decision-Making / Davide Geneletti.
Description: Newyork : Anthem Press, 2019.
Identifiers: LCCN 2019012574| ISBN 9781783089529 (hardback) |
ISBN 1783089520 (hardback)
Subjects: LCSH: City planning. | Environmental impact analysis. |
Decision making. | BISAC: ARCHITECTURE / Urban & Land Use Planning. |
NATURE / Environmental Conservation & Protection. |
SCIENCE / Environmental Science.
Classification: LCC HT166 .G46143 2019 | DDC 333.71/4–dc23
LC record available at https://lccn.loc.gov/2019012574

ISBN-13: 978-1-78308-952-9 (Hbk)
ISBN-10: 1-78308-952-0 (Hbk)

This title is also available as an e-book.

CONTENTS

FIGURES

TABLES

ACKNOWLEDGEMENTS

The research presented in Chapter 2, Chapter 4 and Chapter 5 has been conducted jointly with Dr Blal Adem Esmail (University of Trento), Dr Iris van Duren (University of Twente) and Dr. Francesco Orsi (University of Trento), respectively. I am extremely grateful to them for contributing to this book. Giorgio Menapace provided support to the GIS analysis presented in Chapter 4 and Chapter 6. Finally, I would like to thank Maria Susana Orta Ortiz for helping me with the layout and some of the figures.

AUTHOR AND CONTRIBUTORS

Author

Davide Geneletti is Associate Professor of Spatial Planning at the University of Trento, leader of the Planning for Ecosystem Services research group (www.planningfores.com), former research fellow at Harvard University's Sustainability Science Program and visiting scholar at Stanford University's Woods Institute for the Environment. He has consulted for the United Nations Environment Programme (UNEP), UN-HABITAT and the European Commission. He is also the editor-in-chief of the *Journal of Environmental Assessment Policy and Management* and deputy editor of the open-access journal *One Ecosystem*.

Contributors

Blal Adem Esmail is a postdoctoral research fellow at the University of Trento and member of the Planning for Ecosystem Services (PLANES) research group (www.planningfores.com). His research focuses on applying ecosystem services and nature-based solutions to improve real-life decisions, including project design, planning and policy-making. He worked on operationalizing ecosystem services for watershed management and planning.

Iris van Duren is with the Faculty of Geo-Information Science and Earth Observation of the University of Twente. She is involved in education and research in a wide range of topics: from ecosystem functioning of wetlands via restoration of chimpanzee habitat to forest carbon studies and bioenergy assessments. Her expertise is remote sensing and spatial analysis for various applications in natural resources management.

Francesco Orsi is a research scientist in the Department of Civil, Environmental and Mechanical Engineering at the University of Trento (Italy) and an adjunct professor of geography at Kansas State University (USA), where he taught between 2015 and 2017. His research explores issues and opportunities related to enhancing people's access to nature, focusing on greener landscape planning, human preference towards nature and sustainable transport.

INTRODUCTION

Environmental decisions involve biophysical, sociocultural and economic issues, whose understanding necessitates appropriate methods of knowledge synthesis and stakeholder engagement, as well as the capability to handle increasingly complex and large data sets. It can be argued that all environmental decision-making processes need to consider and compare some forms of alternatives. Alternatives include all types of options, choices or courses of action to accomplish particular goals (Steinemann 2001). These alternatives can be related, for example, to 'different possible purposes, different locations and design, different general approaches to serving the selected purpose, different locations and designs, different packages of mitigation and enhancement components, ad different implementation plans' (Gibson et al. 2005, 126). In environmental decisions, alternatives are assessed against multiple and competing goals, such as protecting nature, addressing needs of communities and fostering economic growth.

Multicriteria analysis (MCA) is an effective knowledge synthesis method that supports decision-making by systematically exploring the pros and cons of different alternatives (Linkov and Moberg 2012; Geneletti and Ferretti 2015), and unveiling trade-offs. It allows comparing alternatives against a set of explicitly defined criteria that account for the most relevant aspects in a given decision-making process. Operationally, MCA supports structuring decision problems, assessing the performance of alternatives across criteria, exploring trade-offs, formulating a decision and testing its robustness. MCA is particularly useful when reducing a multi-objective problem into a single-objective problem is either unfeasible or undesirable, especially in participatory settings involving diverse stakeholders with diverse objectives (Linkov et al. 2006). Criteria can be incommensurable, and expressed through different units of measurement: monetary values, number of new jobs, biophysical units, qualitative evaluations and so on.

The main strength of MCA is that it allows combining the analytical performance of the alternatives with the preferences and priorities of stakeholders in a transparent and replicable fashion (French et al. 2009). For these reasons,

MCA has been increasingly used in environmental decision-making to support the identification of suitable courses of action, by integrating factual information coming, for example, from field surveys or impact modelling, with value-based information collected through stakeholder engagement (e.g. Nordström et al. 2011). Thus, MCA is often integrated with other methods and tools, such as participatory approaches and geographic information systems (GIS), as exemplified in Mustajoki et al. (2011) and Janssen et al. (2014). Participatory approaches include a wide range of methods that aim at eliciting the values, preferences and knowledge of stakeholders, such as Delphi surveys (Bali et al. 2015), focus groups (Comino et al. 2014) and workshops and meetings (Jalilova et al. 2012; Zhang et al. 2013).

The aim of this book is to provide an overview of the principles of MCA, and a series of case studies that illustrate its application to a variety of environmental decision-making problems, ranging from the siting of facilities with critical environmental effects to Natural Park planning, and from the prioritization of environmental restoration interventions to the assessment of the impact of tourism infrastructures. The book is structured as follows. Chapter 1 introduces the principles of MCA and describes the main stages of a generalized MCA process, by providing details and references to support the implementation of each stage. Chapter 2 reviews the application of MCA for a specific field of environmental decision-making: nature conservation. The objective is twofold: to take stock of past experiences by investigating how key stages of the MCA process have been performed, and to compare findings with best practices in order to provide recommendations for successful applications.

In Chapter 3, a case study about landfill site selection is presented. The method is based on the combination of stakeholder analysis and spatial MCA to first design possible sites for a landfill and then rank them according to their suitability. Chapter 4 presents an application of MCA to support protected area planning. The case study illustrates the process of proposing a zoning scheme for a natural park, by combining MCA and multi-objective evaluation. Chapter 5 addresses the problem of forest landscapes restoration. In this case study, spatial and non-spatial MCA are applied to first identify forest reforestation priority areas, and then design landscape-scale reforestation options aimed at improving both ecosystem quality and human living conditions. The last case study, described in Chapter 6, shows how MCA can be combined with GIS-based indicators to assess and compare the environmental impacts of proposed ski areas in a mountain watershed. Finally, the Conclusion provides some findings about the potential of MCA to support environmental decision-making, and the set of skills required for successful MCA applications.

One final note about terminology: MCA is also referred to as multicriteria decision analysis (e.g. in Belton and Stewart 2002) or multicriteria evaluation (e.g. in the seminal work of Voogd (1983)). Although these terms are not strictly synonymous, their use and definition are not always consistent in the literature. For this book, I preferred MCA (as, e.g. in Beinat and Nijkamp 1998) because it has somehow a broader meaning, suggesting that the approach is useful to frame and better understand a decision problem and engage stakeholders and explore their views, as much as to take decisions and select alternatives.

Chapter 1

PRINCIPLES OF MULTICRITERIA ANALYSIS*

1.1 Introduction

Essentially, multicriteria analysis (MCA) is a method to support decision-making, by exploring the balance between the pros and cons of different alternatives. The comparison of alternatives is based upon a set of explicitly formulated criteria, which represent aspects of the alternatives that need to be taken into account during decision-making. More specifically, MCA assists the decision-making process by framing decision problems, illustrating the performance of the alternatives across all criteria, exploring trade-offs, formulating a decision and testing its robustness. All this while considering, on the one hand, the 'analytical' performance of the alternatives across all criteria, and, on the other hand, the preferences and opinions of the stakeholders involved in the process.

Many MCA approaches exist, which differ, inter alia, in terms of computational complexity, level of stakeholder engagement, and time and data requirement. Belton and Stewart (2002) present a detailed analysis of the theoretical foundations of different MCA approaches, as well as their comparative strengths and weaknesses. It could be argued that most approaches share the same main stages of the generalized MCA process shown in Figure 1.1 (based on Kiker et al. 2005; Geneletti and Ferretti 2015), in which a key aspect is the involvement of different actors providing different types of input to the process. The first stage of the MCA process aims at establishing a shared understanding of the decision context, and structuring the problem. Operationally, it includes defining the objectives of the decision process, identifying possible alternatives to achieve them and formulating explicit criteria to assess how each alternative contributes to achieving the objectives. This stage is explained in detail in Section 1.2.

* This chapter draws on sections 11.1–11.3 (solely written by Davide Geneletti) of: D. Geneletti and V. Ferretti, '11. Multicriteria Analysis for Sustainability Assessment: Concepts and Case Studies', in A. Morrison-Saunders, J. Pope and A. Bond (eds), *Handbook of Sustainability Assessment* (Cheltenham, UK; and Northampton, MA: Edgar Elgar Publishing, 2015), pp. 235–64.

The second stage is the actual analysis, broadly consisting of: criteria assessment, weighting, criteria aggregation and sensitivity analysis (see Mundaa et al. 1995; Roy 1996; Belton and Stewart 2002). In short, criteria assessment is the quantification of the performance of each alternative against each criterion defined in the first stage. Weighting accounts for the different levels of importance of the criteria to the decision, and it refers to quantified preferences among possible outcomes for criteria from the perspective of decision-makers and/or stakeholders. Criteria aggregation refers to the application of a rule (e.g. an algebraic expression) to combine the outcome of criteria assessment and weighting and assess the overall performance of each alternative. Finally, sensitivity analysis explores the relationship between the output and the input of the process (Saltelli et al. 2000a,b) and tests the robustness of the results considering the uncertainty factors related to all the previous steps. This stage is explained in detail in Section 1.3.

The third stage of the MCA process is where information from the previous stages is brought together and used to support the decision-making process. In some situations, this stage includes the actual identification of the most suitable alternative(s). In others, it can be limited to producing (and visualizing) information about, for example, the performance of the alternatives associated with different perspectives (i.e. different values or preferences); the contribution of the different criteria to the ranking of alternatives; the exploration of the implication of alliances/convergence between different stakeholders and so on. The case studies presented in the remaining chapters of this book offer insights on the different ways in which MCA results can be used to support decision-making. The objective of this chapter is to expand on the principles of MCA, by fleshing out the content of the first two stages. In particular, Section 1.2 addresses the identification of the decision context and the structuring of the problem, and Section 1.3 describes the subsequent analysis required to run the MCA and produce results.

1.2 Establishing the Decision Context and Structuring the Problem

This first stage aims at establishing a shared understanding of the decision context, including the administrative, political and social structures that surround the decision being made (DCLG 2009). Particularly important is the identification of the final purpose of the decision exercise, and of the actors involved in it. Decision processes may be aimed, for example, at identifying a single most preferred solution, producing a ranking or shortlist of possible solutions, or distinguishing between acceptable and unacceptable solutions. In some cases, decision processes aim at performing a rapid screening of potentially suitable solutions out of a large set, followed by a more detailed assessment of the

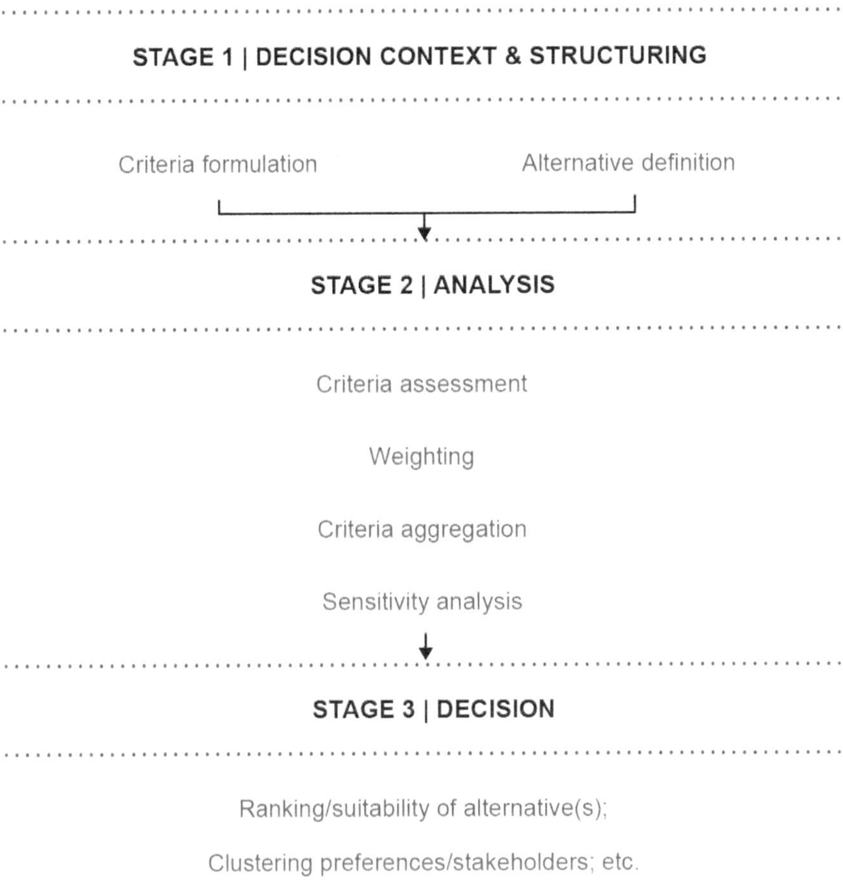

Figure 1.1 A generalized scheme of the mains steps of an MCA process (after Adem Esmail and Geneletti 2018).

selected solutions. A typical example of these stepwise processes is represented by site selection problems (Geneletti 2010).

As to the actors, the following groups of people involved in the decision process need to be identified: decision-makers, stakeholders and experts (French et al. 2009). Decision-makers are those responsible for making the decision. In environmental decision-making, they are often represented by the public authority appointed to formally take the final approval decision about a development option. Stakeholders represent the groups of people that have an interest in and/or that are affected by the outcome of the decision. Hence, they have a claim that their values and opinion should be considered. Evidence suggests that involvement of stakeholders leads to better quality decisions, by adding new information, ideas and analysis (Beierle 2002). Stakeholders are typically identified on a case-by-case basis, even though the legislation may dictate the involvement of specific stakeholder groups. Experts provide professional advice (e.g. scientific, economic, etc.) on the content of the decision, for example, by proposing methods to collect and process data, or by suggesting appropriate criteria and ways to measure them (Orsi et al. 2011). Other important elements of the decision context that need to be understood at this stage are the timeframe of the process and the requirements that may be posed by existing regulations or legislation (e.g. in terms of mandatory consultation processes or mandatory data and indicators).

Once the general context of the problem is clear, the actual structuring (or 'framing') can take place. Structuring consists in identifying the overall and specific objectives of the problem, the possible alternative ways to achieve them and the criteria that can be used to measure the extent to which the different alternatives meet the objectives. Typically, problem structuring is conducted in an iterative fashion, by progressively refining the definition of objectives, criteria and alternatives. Objectives and criteria are often arranged in a hierarchical structure (often called value or decision tree). In this structure, criteria are grouped into sets that relate to separate objectives or sub-objectives, which may refer to different thematic areas (e.g. environmental issues, social issues, etc.). Table 1.1 presents an overview of desirable properties that a set of objectives and criteria should have in order to be used in MCA.

Structuring is a learning process to stimulate discussion and advance the understanding of the problem, including the values and priorities at stake (Costa and Oliveira 2002). Early involvement of stakeholders is instrumental to effective problem structuring, and contributes to increase the acceptance of the final outcome of the analysis (Gamper and Turcanu 2007). One of the most influential pieces of research about problem structuring is the value-focused approach developed by Keeney (1992), which illustrates the creative processes that should be followed to identify and describe the decision problem, articulate the value trees and create relevant alternatives. This approach puts values at the centre of decision-making, claiming that having explicit values makes it easier to define objectives, generate possible alternatives, compare them, and communicate and negotiate the results. As such, it is more proactive and can be particularly helpful in environmental decisions, where there is frequently the need to contribute to the design of possible alternatives, rather than reacting to alternatives already proposed or anyhow readily available (Geneletti 2014).

Problem structuring is a fundamental – though often overlooked – stage, because the way in which a problem is presented to the actors has profound effects on the preferences expressed, and ultimately on the final decision. This

Table 1.1 Desired features of a set of objectives and criteria

Complete	Captures all the fundamental aspects that matter to assess the consequences of the alternatives and compare them (e.g. the key environmental, social, economic, cultural and health effects)
Concise	Uses the smallest number of objectives and criteria to avoid redundancy and double counting. Unnecessary elements (e.g. criteria that appear unimportant or that duplicate existing information) are omitted
Operational	Requires information that can be reasonably acquired (with the required quality), considering the constraint of the decision context
Independent	Contains objectives and criteria that are 'preferentially independent' (Keeney 1992). This means that the performance of an alternative against a given criterion can be assessed without knowing its performance against any other criterion. If preference independence is violated, some MCA approaches run into errors of logic when evaluating alternatives (see Keeney (1992) and DCLG (2009) for more on this)
Understandable	Speaks directly to all the actors involved in the problem, using commonly understood terms and unambiguous language. This may require breaking down general objectives that can be interpreted in different ways (e.g. improve health) into sub-objectives with a clearer meaning (e.g. reduce people's exposure to air pollution), and/or using specifically defined criteria (e.g. number of people living in areas that do not meet a given air quality standard)

Source: Modified after Keeney (1992) and Gregory et al. (2012).

has been widely discussed by the literature in cognitive psychology, after the seminal work of Tversky and Kahneman (1981). For example, the way in which criteria are presented and arranged in a hierarchy, the complexity of the hierarchy (how many levels and how many criteria per level) and the level of 'balance' among the different levels are all issues that should be carefully considered because they have an effect on stakeholders' perceptions about the problem and on their capability to actually deal with the problem (see, e.g., Weber et al. 1988; Keeney 2002). An overview of these issues specific to decision problems in environmental impact assessment can be found in Retief et al. (2013). There are no formal guidelines to dictate how a problem should be structured, and for most environmental problems an unambiguously correct structure does not exist. However, a good structure could be simply defined as 'one that reflects a clear, logical and shared point of view about how the many [objectives and] criteria that may be relevant to an MCA assessment can be brought together into coherent groups, each of which addresses a single component of the overall problem' (DCLG 2009, 34).

1.3 Running the MCA

Many MCA approaches exist (see, e.g., Mundaa et al. 1995; Roy 1996; Belton and Stewart 2002), which differ, among other things, in terms of computational complexity, level of stakeholder engagement, time and data requirements (see DCLG (2009) for a clear and concise overview of the most common MCA approaches). However, after having established the decision context and structured the problem, most approaches involve the following main stages: criteria assessment, weighting, criteria aggregation and sensitivity analysis.

1.3.1 Criteria assessment

MCA does not put any constraint on the type of criteria that can be included in the analysis, as long as they are clearly linked to the objectives and are meaningful for the actors involved in the exercise. So, criteria can be expressed in quantitative units, in qualitative terms, using a linguistic scale, a Boolean variable and so on. Once the value tree is laid out and the alternative identified, one needs to populate the resulting decision matrix (see example in Table 1.2), by assessing the performance of each alternative against each criterion. Conceptually, this requires two steps (Geneletti 2005a). The first step consists in providing the 'raw' information about the effects of the alternatives on the different criteria. In many environmental decision-making processes (e.g. environmental impact assessment, see Chapter 6) this information usually comes from impact analyses, conducted through modelling, measurements, fieldwork and so on. For example, the analysis may predict that one alternative causes an average daily noise level of, say, 65 decibel.

The second step consists in converting this information into a preference scale, that is, a dimensionless expression of the level of desirability of the alternative. Through this conversion, the original scores lose their natural units of measurement and are converted into a consistent numerical scale. For the sake of simplicity, typically this scale ranges between zero and one, where zero corresponds to the minimum level of preference that is likely to be encountered given the decision context, and one to the maximum level. So, for example, the 65 noise level may become a zero, because, say, it exceeds a given quality threshold and corresponds to an unacceptable performance (considering the specific context of the decision process). The conversion requires input from relevant actors (experts and/or stakeholders, depending on the criterion under consideration), which can be provided by using specific techniques, such as, for example, value functions (Figure 1.2). Value functions represent mathematical representations of human judgement (Beinat 1997), and their use in MCA for

Table 1.2 Example of decision matrix for a hypothetical hotel development, where four alternative sites are compared across five different criteria

	Site A	Site B	Site C	Site D
Proximity to the beach (m)	50	0	150	500
Landscape beauty (–/++)	–	+	–	++
Slope (%)	5	2	25	15
Real estate cost (1,000$/ha)	64	100	13	24
Golf field within 2 km (Y/N)	N	N	Y	Y

Source: Geneletti and Ferretti (2015).

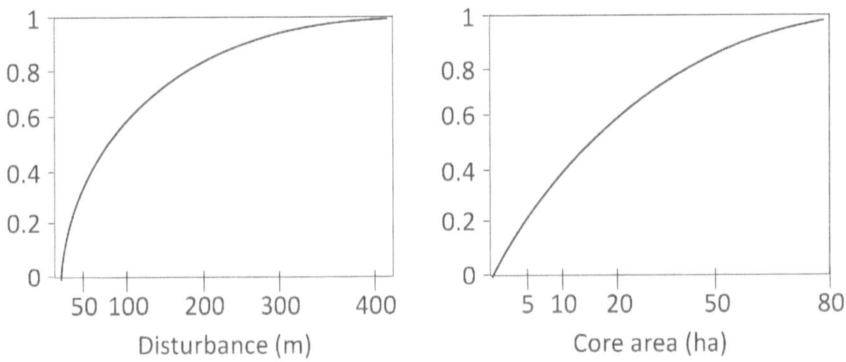

Figure 1.2 Examples of value functions for two criteria (exposure to disturbance factors and core area) used to compare the nature-conservation relevance of forest patches in a human-dominated landscape. The original criterion values are indicated on the x-axis, and their normalized value (on a 0–1 scale) on the y-axis. *Source*: Geneletti (2004b).

environmental problems is well documented in the literature (Geneletti 2004a; 2005a,b). At the end of the criteria assessment stage, a normalized decision matrix is produced, where all criteria are expressed in the same dimensionless value range.

There are techniques, such as the analytic hierarchy process (AHP, Saaty 1980), that approach the assessment of criteria scores indirectly, by eliciting from the actors a series of qualitative judgements that are then used to construct the normalized matrix. In AHP, both criteria scores and weights are derived through questions that ask decision-makers to compare pairs of criteria and alternatives, respectively. AHP is one of the most popular MCA approach, widely discussed in the scientific literature (e.g. Saaty and Vargas 2012; Bhushan and Rai 2004). A comprehensive overview of how different

MCA techniques address criteria assessment can be found in Belton and Stewart (2002) and DCLG (2009).

1.3.2 Criteria weighting

In most decision problems, criteria have different priority levels. Hence, the values in the normalized matrix cannot be directly combined because a unit of preference of one criterion is not necessarily equal to a unit of preference of another criterion. MCA uses the concept of 'weight' to account for this during the analysis. A weight is a value assigned to a criterion that indicates its importance relative to the other criteria under consideration (Malczewski 1999). Analogously to the preference assessment of criteria described in the previous section, also the selection of weights requires the consultation of the actors involved in the decision-making process. Actors are asked to provide their judgements and perceptions about the importance levels of the different issues at stake, which are then converted into ordinal or cardinal weights, through a number of different techniques (Table 1.3).

A number of methods developed for stakeholder consultation can be used to support weight elicitation in MCA, including Delphi surveys, focus groups, and interviews and questionnaires. Delphi surveys aim at soliciting the advice of a panel of experts, and whenever possible forging a consensus (MacMillan and Marshall 2006). The approach is based on structured and written questionnaires to which panelists are asked to answer anonymously. All responses are summarized and reported back to panelists who have the opportunity to revise their judgements. The use of Delphi surveys to assess weights in MCA has been exemplified in Geneletti (2007a). Finally, the use of interviews to derive sets of weights for different stakeholder groups is presented, for example, in Uribe et al. (2014).

One final technical note. The concept of weight takes on slightly different meanings, and carries different implications, according to the specific MCA method that is used. Particularly, when techniques based on weighted linear aggregation of criteria are used (see next section), weights do not have an absolute or intrinsic meaning, and need to be elicited with reference to the range of variations of the criteria that occur in the specific decision problem (van Herwijnen 1999). In these cases, weights basically express how many units of a given criterion we are willing to sacrifice in order to gain one unit of another criterion. However, this is not the case when techniques based on outranking methods are used (see Section 1.3.3). So, weights need to be handled with care and the correct information about their meaning and implications needs to be conveyed to the actors involved in their elicitation. More details on these aspects for different MCA methods can be found in Choo et al. (1999).

Table 1.3 Overview of some of the most commonly used techniques to determine weights in MCA

Technique	Description	Reference
Trade-off	Based on a set of questions that explicit to what extent an increase of one unit in a given criterion is attractive with respect to similar increases in the other criteria	Keeney and Raiffa (1976)
Analytical hierarchy process	Based on pairwise comparison of criteria that are measured using a predefined ratio scale	Saaty (1980)
Swing	Based on the analysis of the value advantage associated with swinging a given criterion between the endpoints of its range (i.e. from the worst hypothetical effect to the best one)	von Winterfeldt and Edwards (1986)
Random	Uses ordinal information on criteria priorities and a random generator to determine a large number of possible weights, uniformly distributed with the set of feasible weights. Results are used to assign probabilities that a given alternative will rank in a certain position	Janssen et al. (2001)
Extreme weights	Uses ordinal information on criteria priorities and generates the different combinations of extreme quantitative weights that fulfil the ordinal information	Janssen et al. (2001)
Expected value	Uses ordinal information on criteria priorities and generates the expected value within this ordering to calculate quantitative weights	Janssen et al. (2001)

Source: Geneletti and Ferretti (2015).

1.3.3 Criteria aggregation

During criteria aggregation, a decision rule is applied to combine the performance of the alternatives across all criteria, by using the information generated in the previous two stages. This produces a ranking of the alternatives. Different MCA methods may differ in the decision rule that they apply. The simplest, and possibly the most widely used, decision rule is the weighted linear combination (WLC). WLC is based on the concept of weighted average, and produces an overall score for each alternative by multiplying the normalized criterion scores by their weight, and then summing the products over all criteria (Malczewski 1999). The use of WLC is appealing because the method is straightforward, and it is easy both to explain and to compute. However, WLC is appropriate only if the criteria are mutually preference independent (see Table 1.1). Additionally, WLC is a compensatory method. That is, poor achievements on some criteria can be compensated by better achievement on other criteria. If this does not apply to a specific decision context, one may prefer to use different methods, or introduce thresholds above/below which the compensability is not acceptable.

Other methods for criteria aggregation include the family of outranking methods (e.g. ELECTRE, Roy 1996), the family of ideal/reference point methods (e.g. TOPSIS, Malczewski 1999) and fuzzy methods (e.g. NAIADE, Mundaa et al. 1995). Outranking methods are based on the idea of eliminating alternatives that are 'outranked'. In short, one alternative outranks another if it performs better than the other on enough criteria of sufficient importance (as reflected by the sum of their weights), and it is not outperformed (as reflected by a significantly lower performance) by the other on any one criterion. A comprehensive survey of methods is presented in Yoon and Hwang (1995). Huang et al. (2011) review the applications of the most popular methods to environmental decision-making. Ultimately, the choice of the preferred MCA method should be based on the needs and context of the specific decision-making process, including the foreseen level of interaction with the actors and the available expertise.

1.3.4 Sensitivity analysis

Sensitivity analysis explores the relationships between the output and the inputs of a model (Saltelli et al. 2000a,b). In MCA, sensitivity analysis can be conducted to determine the robustness of the results with respect to uncertainty factors associated with any of the previous stages. The rationale for conducting a thorough sensitivity analysis is the fact that most decisions are affected by uncertainties, owing to measurement and conceptual errors, limited knowledge about process, simplification and data scarcity and, last but not least, different values and opinions. Sensitivity analysis considers how, and how much, such errors and uncertainties affect the final results, that is, the ranking of alternatives. It aims to answer questions such as: How stable is the ranking? Up to what point can the input factors be changed without affecting the results? And ultimately: Can the results be trusted?

Although often overlooked in environmental decision-making (Janssen 2001), sensitivity analysis is a fundamental step that offers a huge added value to an MCA, and to its results. For example, sensitivity analysis can be used to generate several rankings of alternatives under different perspectives (i.e. different weight sets), contributing to highlight the reasons for divergence among stakeholders, and the key issues that require attention during the debate. It emphasizes the role of MCA as a learning (and iterative) process, where the actors progressively improve their understanding about the problem and identify suitable ways forward. Through sensitivity analysis one can learn not only about the stability of the performance of the alternatives, but also about the factors that most contribute to it. Hence, the results can be used to direct further efforts, for example, in terms of data gathering or analysis, involvement of experts of specific areas, collection of additional stakeholders' opinion and so on.

Several techniques exist to perform sensitivity analysis. The most popular ones (that are commonly implemented in MCA software) test the sensitivity with respect to changes in the weights and the scores of the criteria. A way of performing this consists in imposing perturbations on the original values to verify their effect on the overall performance of the alternatives (Janssen et al. 2001). First of all, it is necessary to identify suitable uncertainty ranges for every weight/impact score. Such a range represents the maximum percentage by which we expect the weight/score to deviate from its original value, due to model limitations, data inaccuracy, disagreement among experts and so on. Afterwards, the weights/scores are calculated a large number of times using a random number generator within the predefined uncertainty ranges. The multicriteria aggregation is then repeated for each weight/score set, generating frequency tables that show how often each alternative ranks in each position (examples can be found in Geneletti 2005b; 2010).

Another useful technique consists in studying the sensitivity of the ranking with respect to one of the criterion weights, by computing the rankings corresponding to all possible values for that weight. During this procedure the ratios of the remaining weights are assumed the same as those of the original weights. This analysis is useful to identify the reversal points, that is, the weight values that cause a reversal in the rank order of two (or more) alternative (see example in Figure 6.8). Further sensitivity analysis techniques are described in Janssen et al. (2001) and Delgado and Sendra (2004).

1.4 Conclusions

This chapter illustrates the key features and the main stages of a generalized MCA process, from the establishment of the decision context and the structuring of the decision problem, to the actual analysis and use of its outcomes to support the decision-making process. The analysis can be broken down into a few steps (criteria assessment, weighting, aggregation and sensitivity analysis), which are performed differently, according to the specific MCA approach that is used. The next chapter adopts the same generalized MCA scheme to review applications in the field of nature conservation. The review illustrates how the different steps described in this chapter have been applied in practice, and what are the emerging strengths and common pitfalls.

Chapter 2

A REVIEW OF MCA APPLICATIONS FOR NATURE CONSERVATION*

2.1 Introduction

Decision-making in nature conservation increasingly requires comparing alternatives to achieve multiple and competing goals, such as promoting habitat conservation, while addressing the well-being of vulnerable communities (e.g. Zia et al. 2011; Zhang et al. 2013; Karlson et al. 2016). Alternatives include, for example, different locations and design for natural parks or restoration areas (Van Elegem et al. 2002; Uribe et al. 2014), options for plant species (Reubens et al. 2011) or wildlife management units (García-Marmolejo et al. 2008). For these reasons, MCA has been increasingly used in conservation to support the identification of the most suitable alternative(s), by integrating factual information coming from surveys or modelling, with value-based information collected through stakeholder engagement (e.g. Strager and Rosenberger 2006; Nordström et al. 2011). In order to be useful, MCA, and its integration with other methods and tools, need to be carefully designed and tailored to the specific context and decision-making process. In nature conservation, this has led to a variety of approaches and applications.

This chapter provides a structured review of empirical applications of MCA to nature conservation published in the scientific literature over the past 20 years. It adds to a more general review of MCA applications in environmental sciences carried out by Huang et al. (2011) and its follow-ups based on text mining by Cegan et al. (2017) and Kurth et al. (2017). The review has two objectives. First, to take stock of past experiences by investigating how key stages of the MCA process have been performed. Second, to compare these findings with best practices and common pitfalls identified in the MCA literature, in order to provide recommendations for successful applications in nature conservation, and more generally environmental decision-making.

* The text and graphics of this chapter are based on the following paper: B. Adem Esmail and D. Geneletti, 'Multi-criteria Decision Analysis for Nature Conservation: A Review of 20 Years of Applications', *Methods in Ecology and Evolution* 9 (1) (2018): 42–53.

The selection of MCA applications was conducted through keyword search in the Scopus database (www.scopus.com) limited to studies published between 1996 and 2016 that contain nature or biodiversity conservation and multicriteria analysis (or relevant acronyms and synonyms, including multicriteria decision analysis, evaluation, etc.) in the title, abstract or keyword. The search resulted in a total of 774 articles. After excluding review articles and non-relevant articles, the sample was reduced to 86 articles, which reported actual MCA applications in nature or biodiversity conservation. The list of selected articles is presented in the Annex.

The review of the selected articles was performed by developing a review framework based on the main stages of the generalized MCA process presented in Chapter 1 (see Figure 1.1). The framework consists in a set of questions, and associated interpretation keys (Table 2.1). The framework was applied to each MCA application reviewed, first, by answering the questions with text extracts and notes from the article; and then classifying this information using the interpretation keys. The results, presented in Section 2.2, shed light on how the literature addressed the three main stages of the MCA process, describing the techniques that have been used and the level of involvement of different actors. In Section 2.3, the findings are discussed in the light of best practices and common pitfalls reported in the MCA literature, to provide recommendations to future applications.

2.2 Findings of the Review

On average, six articles have been published yearly since 2002. As shown in Figure 2.1, the case studies were located in Europe (49 per cent), Asia (17 per cent), North America (13 per cent), Africa (9 per cent), South America (6 per cent) and Australia/New Zealand (6 per cent). In total, the articles covered 41 countries; the most represented countries being Italy (thirteen articles), the United Kingdom (seven articles), the United States (six articles), followed by Iran (five articles), Australia, Mexico and Spain (four articles each), and Finland, France and Sweden (three articles each) and the remaining 31 countries covered by one or two articles. In the sections that follow the articles are referenced within square brackets according to their numbering in the Annex.

2.2.1 Decision context and structuring

Overall, the reviewed articles covered five broad and not mutually exclusive topics. Most of the articles addressed forest management and restoration (25 per cent), conservation prioritization and planning (24 per cent) and protected area planning and management (21 per cent), while fewer dealt with mapping of biodiversity, wilderness and naturalness (8 per cent). Finally, the remaining

Table 2.1 The framework adopted to review MCA applications

Question	Interpretation key
STAGE 1: Decision context and structuring	
Geographical region of application and publication year	Country in which the study area is located and year of publication
What is the objective of the decision-making process? What is the spatial extent of analysis?	Main objective of the MCA process described (it may be different from the overall objective of the article) Spatial extent of the study area, i.e., less than 500 km^2, between 500 km^2 and 250,000 km^2, between 250,000 km^2 and 1,500,000 km^2, or above 1,500,000 km^2
Are alternatives identified? If yes, how?	0 = No. No predefined alternatives 1 = Yes. Alternatives are already existing 2 = Yes. Alternatives are identified/designed during the study solely by the authors 3 = Yes. Alternatives are defined/designed during the study, with input from stakeholders
How are criteria formulated?	0 = Formulated by the authors 1 = Formulated with input from other experts 2 = Formulated with input from stakeholders (if applicable, specify the engagement technique) 3 = Formulated with input from the general public " (if applicable, specify the engagement technique)
STAGE 2: Analysis	
How are criteria assessed (i.e. how the performance of the alternatives is converted from the original unit of measurement into a dimensionless preference scale)?	0 = Not described /not performed 1 = Assessment performed by the authors without providing justification 2 = Assessment performed by the authors with justification 3 = Assessment performed based on input from experts or stakeholders
How is weighting performed?	0 = Not described /not performed 1 = Weighting performed by the authors 2 = Weighting performed by the authors through multiple weight sets that simulate different perspectives 3 = Weighting performed by involving experts or representatives of stakeholders 4 = Multiple weighting performed by involving experts or representatives of stakeholders
Criteria aggregation	Decision rule applied to aggregate the performance of the alternatives across criteria

(*continued*)

Table 2.1 *(cont.)*

Question	Interpretation key
Sensitivity analysis	0 = Not mentioned/nor performed 1 = By changing weights (if applicable, specify the technique adopted) 2 = By changing scores 3 = By changing criteria assessment method 4 = By changing criteria aggregation rule 5 = Testing the robustness of the results by collecting general feedback from participants 6 = Testing the robustness of the results by consulting additional data sources
STAGE 3: Decision	
What information is provided by the MCA to support the final decision? (more answers possible)	0 = Overall alternative ranking or suitability 1 = Partial ranking or suitability according to specific criterion included 2 = Multiple ranking or suitability according to different perspectives 3 = Analysis and clustering of preference/stakeholders

22 per cent of the articles addressed a diverse set of topics, ranging from siting of solar energy development in areas with low biodiversity value to designing railroad corridors minimizing biodiversity impacts. In terms of geographic extent, 56 per cent of the case studies covered an area between 500 km^2 and 250,000 km^2 (i.e. regional extent) and 33 per cent covered an area of less than 500 km^2 (i.e. local extent). Fewer studies, 7 per cent and 5 per cent, were carried out at national (between 250,000 km^2 and 1,500,000 km^2) and continental levels (above 1,500,000 km^2), respectively.

Concerning the definition of alternatives, 41 per cent of the reviewed MCA applications had no predefined alternatives (see Figure 2.2); this includes studies aimed at selecting sites of biodiversity value suitable for protection or restoration [e.g. 41; 71], supporting zoning of terrestrial and marine protected areas [e.g. 8; 23] and mapping biodiversity assets or habitat suitability [e.g. 77; 72]. About 24 per cent considered already existing alternatives. For example, [26] prioritized 104 existing protected areas (i.e. Natura 2000 sites) based on their significance for biodiversity; [9] analysed 744 vacant parcels in New York City to select those that could cost-effectively meet conservation goals; [43] considered 5 alternative land uses to compare their provision of ecosystem services. In 20 per cent of the MCA, alternatives were designed by the authors, applying: spatial MCA to identify restoration sites [e.g. 45; 41], heuristic algorithms to identify additional conservation sites [e.g. 18; 20] or forest modelling to simulate different management scenario

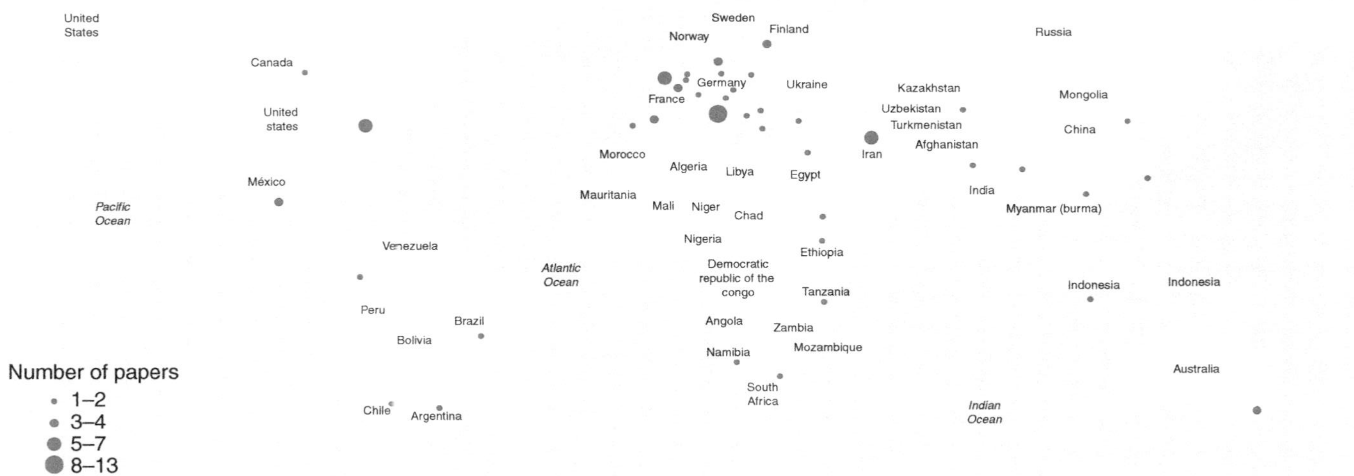

Figure 2.1 MCA applications described in the reviewed articles grouped by country (dots placed on the capital cities).

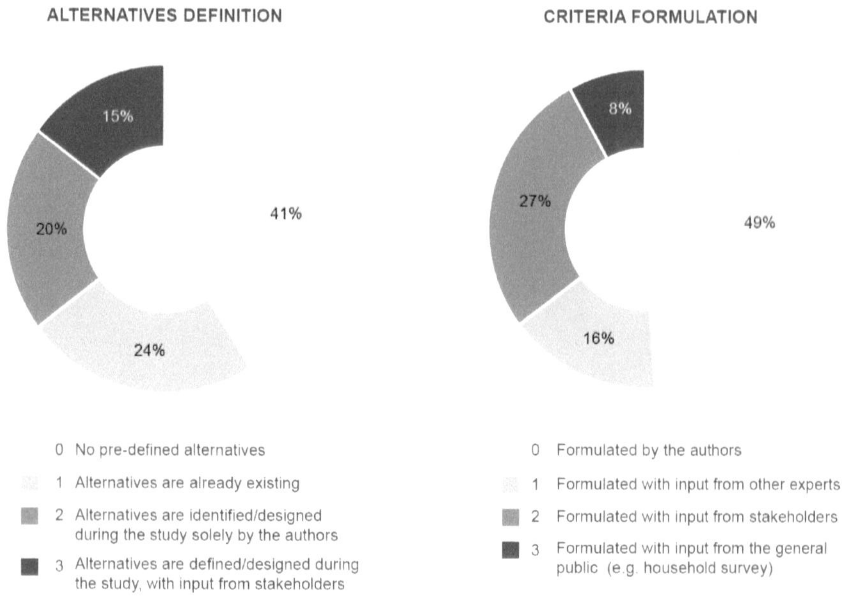

Figure 2.2 Identification of alternatives and formulation of criteria in the reviewed articles.

[e.g. 56; 57; 58]. Finally, in 15 per cent of the cases, the alternatives were defined together with stakeholders; for example, [52] involved local communities to identify woody species for restoration in Ethiopia, [24] defined protected area management scenarios in a participatory fashion in Greece, while [54] involved stakeholders in identifying potential areas for forest restoration in Argentina.

About criteria selection, in 49 per cent of the MCA the authors formulated the criteria without input from stakeholders, mostly relying on literature reviews to justify the selection of the criteria. In fewer cases, reference to legislation was made, such as the European Union Directives on Habitats, Birds and Strategic Environmental Assessment [2; 18], national action plans [59] or previous projects [40]. In 16 per cent of the MCA, experts other than the authors were involved, for example, through Delphi techniques [e.g. 25], focus groups [e.g. 73] and face-to-face meetings [e.g. 51]. In 27 per cent of the MCA, stakeholders' representatives were also involved, mainly thorough workshops [e.g. 13; 24; 28; 40; 46; 50; 58], interviews [e.g. 22; 29; 36] and focus groups [e.g. 26; 80; 83]. Finally, 8 per cent of the MCA included larger groups of stakeholders, for example, through workshops [e.g. 27; 48], participatory meetings [e.g. 17; 30], consultations and open meetings [e.g. 53; 62] and

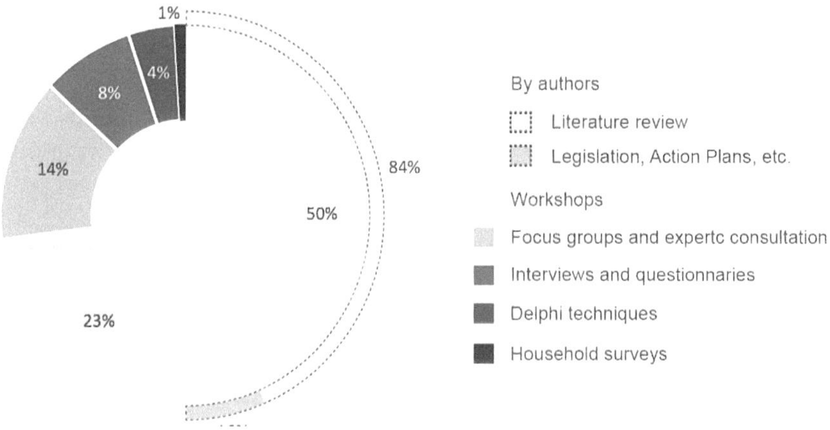

Figure 2.3 Participatory technique and justification of the formulation of the criteria in the reviewed articles.

household surveys [6]. The participatory techniques used to involve experts and stakeholders in the criteria formulation are illustrated in Figure 2.3.

2.2.2 Analysis

Concerning criteria assessment, in 21 per cent of the articles the authors performed it (e.g. by applying mathematical functions or direct assessment) without providing any further justification of the approach [e.g. 50; 75], whereas in 47 per cent a justification was also included [e.g. 46; 49]. About 22 per cent involved stakeholders in defining the criteria assessment method [e.g. 53; 58]. On the other hand, 10 per cent of the reviewed MCA partially overlooked the criteria assessment by avoiding the step related to the conversion into a dimensionless preference scale [e.g. 29; 40].

Weighting was not considered by 2 per cent of the MCA [e.g. 41; 77], while in 26 per cent and 17 per cent the authors themselves assigned a single [e.g. 20; 81] or multiple [e.g. 13; 63] sets of criteria weights, respectively. In 24 per cent and 17 per cent of the MCA applications, stakeholders were involved in assigning a single [e.g. 25; 65] or multiple [e.g. 28; 36] sets of criteria weights. About criteria aggregation, the weighted linear combination was the most widely applied approach (63 per cent), followed by analytic hierarchic/network process (AHP/ANP) [e.g. 48; 40] and the Multi-attribute value theory [e.g. 47; 78]. Other approaches included the Novel Approach to Imprecise Assessment and Decision Environments (NAIADE) [e.g. 24; 32], Ordered Weighted Average (OWA) [e.g. 1; 10], PROMETHEE [e.g. 27; 46] and

SMART [e.g. 52]. In 7 per cent of the cases, more aggregation methods were combined [e.g. 82], and finally, 5 per cent of the reviewed articles applied some ad hoc rules [e.g. 81].

Surprisingly, of the reviewed articles, 43 per cent did not report about any sensitivity analysis. Among the others, the sensitivity analysis was mostly applied to the weights (19 per cent) [e.g. 56; 76], and only in fewer cases the scores (6 per cent) [e.g. 28; 73], the aggregation rule (3 per cent) [e.g. 32; 82] and the assessment method (1 per cent) [11]. In 15 per cent of the MCA, it was performed by changing two factors of the above [13; 60] and in 5 per cent three factors [70; 80]. Finally, in 5 per cent of the articles the robustness of the results was tested by collecting feedback from participants on the MCA results [e.g. 14; 79], and in 3 per cent by consulting additional data sources [e.g. 68; 86]. Figure 2.4 summarizes the analysis conducted in the reviewed articles.

2.2.3 Decision

Almost half of the reviewed MCA applications only provided an overall ranking of the alternatives or an overall suitability [e.g. 11; 12]; while 22 per cent also included partial rankings or suitability [e.g. 2; 5]. Fewer articles (15 per cent) reported multiple rankings, reflecting different perspectives [e.g. 13; 44], and 8 per cent performed analysis and clustering of the preferences of stakeholders [e.g. 47; 68]. Only few articles included more information to support decision-making. For example, [73; 78] provided multiple overall and partial suitability results; [86] presented both partial rankings as well as an analysis of the preference of the stakeholders; finally, [36] reported multiple rankings, together with an analysis of the different stakeholders' preferences. Figure 2.5 shows the analyses of the type of information provided in the articles to support decision-making.

2.3 Discussion and Recommendations for Best Practices

In the following subsection, the findings are discussed in the light of MCA best practices and common pitfalls, providing recommendations for more effective applications in nature conservation, and environmental decision-making in general.

2.3.1 Decision context and structuring

Structuring should be an iterative learning process that aims to enhance the understanding of the problem, and recognize the values and preferences at play (Bana e Costa and Oliveira 2002). Mustajoki et al. (2011) and Ferretti

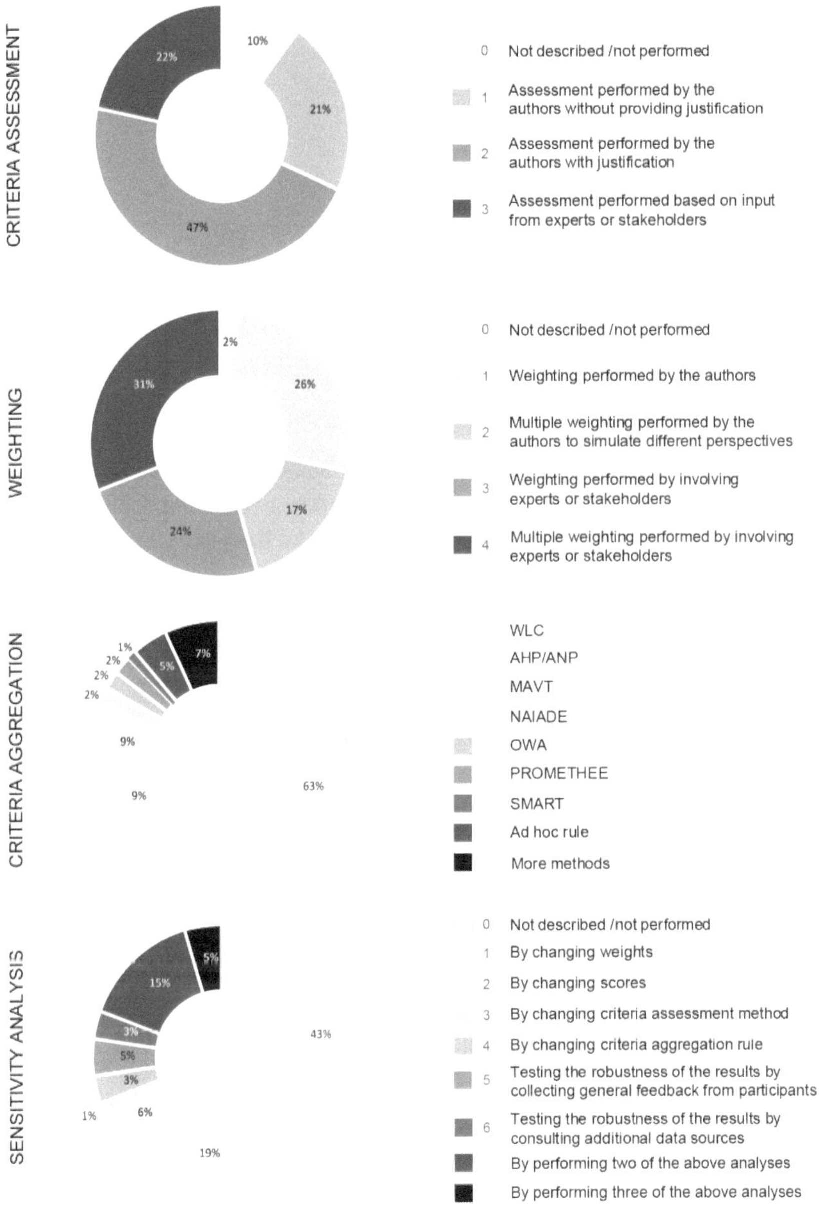

Figure 2.4 MCA analysis conducted in the reviewed articles.

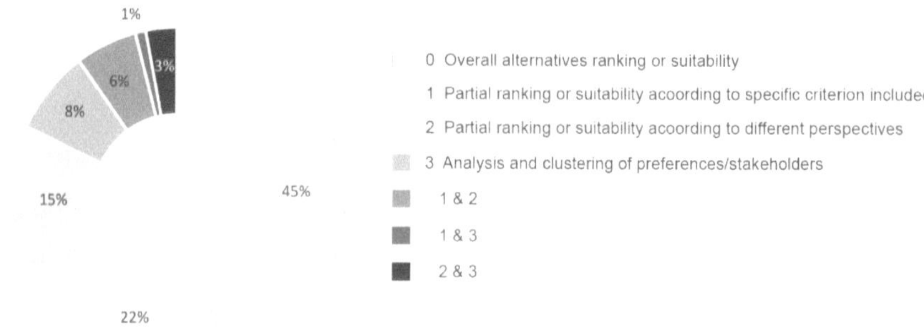

Figure 2.5 Classification of the articles based on the type of information provided to support decision-making.

and Pomarico (2013), for example, refer to Keeney's (1992) value focused approach, illustrating the creative process to define a decision problem, articulate value trees and create relevant alternatives. As Keeney (1992) puts it, the values of stakeholders should be at the core of the decision-making, given that having explicit values makes it easier to define objectives, generate possible alternatives, compare them, and communicate and negotiate the results. Achieving a meaningful participation of the key actors is an important aspect, crucial also for increasing the acceptance of the final outcome of the MCA (Gamper and Turcanu 2007).

Nevertheless, only a relatively small percentage of the reviewed articles engaged stakeholders in identifying alternatives and formulating criteria (15 per cent and 35 per cent, respectively). This confirms that, also in conservation decision-making, limited or poor problem structuring remains a common pitfall of MCA applications, despite the fact that it has been shown to affect stakeholders' preferences, and ultimately the final decision (e.g. Tversky and Kahneman 1981). The hierarchical arrangement of criteria deserves special attention, given that it affects stakeholders' perception of the problem, and their capability to actually deal with it (see, e.g., Weber et al. 1988; Retief et al. 2013).

Objectives and criteria must be chosen to fit the problem according to stakeholders' concerns rather than conform to any standardized expert-based hierarchy or criteria. Good problem structures avoid excessive and unbalanced number criteria for different objectives. Moreover, it is important that a representative set of alternatives is considered, which includes all alternatives that are relevant for the stakeholders (Janssen et al. 2014). More in general, prior to the selection of objectives and criteria, there ought to be a problem definition step aimed at identifying a decision question along with relevant decision-makers, stakeholders and topic experts, ideally facilitated by an expert on participatory decision analysis (see, e.g., Thorne et al. 2015).

2.3.2 Analysis

Criteria assessment is a key phase in a successful MCA. The results show that in most MCA applications (47 per cent), the authors properly justified the step of criteria assessment, often by involving other stakeholders (22 per cent). Ferretti and Comino (2015) is a good example of involvement of diverse actors (i.e. park authorities, residents, tourist, entrepreneurs and researcher) to formulate and validate value functions through focus groups. However, in some of the studies (e.g. Lu et al. 2014), criteria assessment was partially overlooked, for example, by considering only measurable, quantitative criteria (e.g. distance, extension, etc.), but without any further conversion into a dimensionless preference scale.

Weighting was well addressed in the reviewed MCA applications: the majority of the articles considered either single or multiple sets of weights, often defined together with stakeholders; only in few articles (2 per cent) was weighting overlooked. Moreover, as compared to other steps of the MCA process, weight elicitation emerged as the step in which stakeholders' preferences were included more straightforwardly. For example, to derive weights, Ianni and Geneletti (2010) applied a three-point scale pairwise comparison, which means that each criterion can be equally important, slightly more important or strongly more important than the criterion it is being compared to. This simplified approach was considered more appropriate than the more traditional seven- or nine-point scales because of the mostly qualitative nature of the criteria and the varied fields of expertise of the assessors (e.g. indigenous community members).

A good practice is to test the consistency of the judgements expressed by stakeholders involved in the assignment of weights. This could be done by calculating a consistency ratio to measure how consistent the judgements have been relative to large samples of purely random judgements (Saaty 1980). Accordingly, if the ratio is much in excess of 0.1, the judgements are untrustworthy given they are too close for comfort to randomness and the exercise is valueless or must be repeated. In the sample of studies, consistency checks on weights was performed in only 14 per cent of the cases. Among these, Schwenk et al. (2012) argued that checking the consistency ratio of the respondents helped stakeholders to discuss and redefine preferences. Habtemariam and Fang (2016), on the other hand, arguably accepted a lower consistency threshold (0.2) of stakeholders from various ministries involved in zoning of a marine protected area; while Derak and Cortina (2014), stated that they used a software (i.e. PriEsT) to computationally resolve consistency issues involving 38 stakeholders. Finally, Marre et al. (2016) gave respondents a possibility to revise inconsistent judgements.

Criteria aggregation, by combining weights and scores to derive overall values of each alternative across all criteria, is a crucial MCA step leading to the final ranking of the alternatives. The results show that the most applied algorithm was the weighted linear combination, generally preferred because of its straightforwardness, intelligibility by stakeholders and computation ease. Yet, few articles alone explicitly mentioned that weighted linear combination is appropriate only if the criteria are mutually preference independent. Among these, Ferretti and Comino (2015) noted that independence between attributes should be fulfilled, and checked it with participants in a focus group. Similarly, Mustajoki et al. (2011) and Karlson et al. (2016) clearly assumed the adopted criteria to be mutually preferentially independent (as in Keeney and Raiffa 1976). Geneletti (2007b) tested for correlation among criteria, in order to avoid the use of correlated criteria that may cause redundancy and double counting, and generate inconsistent results.

Sensitivity analysis is essential for testing the robustness of the results considering the uncertainty factors related to all the steps of an MCA. Sources of uncertainty in decision processes include measurement and conceptual errors, limited knowledge about process, simplification and data scarcity and, above all, different values and opinions of stakeholders. Sensitivity analysis can be done through a number of techniques (see Section 1.3.4). Nevertheless, sensitivity analysis was overlooked by almost half of the applications in nature conservation. On the contrary, 20 per cent of the MCA applications present good examples of detailed sensitivity analysis, combining several techniques. Two rather unconventional ways of performing sensitivity analysis were identified in the review: collecting feedback on the MCA results from participating actors, and consulting additional data sources.

2.3.2 Decision

MCA ultimately aims to assist people in making decisions related to the objectives of the specific case study. It can serve diverse purposes in supporting decision-making, including providing structure to the debates between key actors, documenting the process of analysing the decision, making value judgements explicit, creating shared understanding and generating a sense of common purpose (DCLG 2009, 111). These diverse purposes clearly emerged when examining individual studies in the review. In Riccioli et al. (2016), for example, the objective of the MCA was to assess the effectiveness of existing environmental policies. To this end, their main results consisted of spatial clusters identifying areas with high biodiversity based on a set of criteria which were then cross-checked to see whether they overlap with existing protected areas in the region. Again, Marre et al. (2016) aimed to explore how different stakeholders value different aspects involving large-scale marine developments. To this end, they co-developed with stakeholders a hierarchical value structure for impact assessment of a hypothetical project, which then they used to explore, through a national survey, information preference of different types of stakeholders. In this latter case, beyond the overall ranking, the most interesting result was the cluster analysis of the preferences of the different types of respondents.

2.4 Conclusions

A few common pitfalls in MCA application emerged from this review of applications in nature conservation. One of these pitfalls is poor problem structuring, due to, for example, ill-defined alternatives, excessive and unbalanced number of criteria for different objectives and limited stakeholder involvement. A successful MCA application should start by clearly establishing the decision context, defining the purpose of the MCA and identifying the main actors to be involved in the process. This is to be followed by a collaborative identification of an inclusive set of alternative options to meet the desired objective.

In the analysis stage, most of the reviewed articles did not explicitly discuss the rationale behind the criteria assessment step. This is another common pitfall of MCA applications, in which little attention is given to how information about performance of each alternative is converted into a dimensionless scale of preference that expresses the level of desirability of that alternative. Adequate understanding about the implication of weights is another crucial aspect of a successful MCA; however, in the reviewed articles, such information was often scarcely conveyed to the actors involved. A successful MCA application should include an adequate description of the performance of each alternative option against each criterion. Furthermore, weights should be assigned to account for the different degrees of importance of the criteria to the decision, considering different stakeholders' perspective. Hence, a rule has to be applied to combine weights and scores in the overall values for each alternative. Examining the results of an MCA in the light of its initial purpose is perhaps the most important step to assist people in making decisions. Finally, a successful MCA application should always include a sensitivity analysis to examine the trustworthiness and robustness of its conclusions.

Chapter 3

LANDFILL SITE SELECTION: COMBINING SPATIAL MCA WITH STAKEHOLDER ANALYSIS*

3.1 Context

3.1.1 Environmental decision-making problem

Landfill site selection is a complex environmental decision-making process that involves the collection and processing of information relating to environmental, socioeconomic, as well as operational aspects. There is a large body of scientific literature that addresses the location of waste disposal areas, and more in general the interaction between planning and waste management (see the comprehensive annotated bibliography presented in Hostovsky (2000)). In the last couple of decades, the combination of geographical information systems (GIS) and MCA has been routinely adopted as an approach to assess the suitability of an area to host a landfill, and consequently to select optimal locations for these facilities (Delgado et al. 2008). This approach takes advantage of both the capability of GIS to manage and process spatial information, and the flexibility of MCA to combine factual information (e.g. soil type, slope, infrastructures) with value-based information (e.g. expert's opinion, quality standards, participatory surveys). Taking into account both technical elements and people's values and perceptions is essential to build consensus around a decision, reduce conflicts and consequently pave the way to successful landfill siting interventions (Higgs 2006; Norese 2006; Petts 2001; Hokkanen and Salminen 1997; Wichelns et al. 1993).

3.1.2 Case study description

The case study concerns the siting of an inert landfill in Trentino, an alpine region in northern Italy. The region hosts about 90 landfill sites of limited

* The text and graphics of this chapter are based on the following paper: D. Geneletti, 'Combining Stakeholder Analysis and Spatial Multicriteria Evaluation to Select and Rank Inert Landfill Sites', *Waste Management* 30 (2) (2010): 328–37.

capacity (less than 5,000–20,000 m^3), and most inert wastes are disposed locally, that is, in the same municipality where they are produced. Regional guidelines concerning waste management were issued in order to comply with the national legislation, which in turn implemented the EU Landfill Directive (1999/31/CE). These guidelines set new and stricter requirements concerning the authorization to operate an inert landfill, environmental criteria for landfill sites, as well as the management of the site. The strategy devised by the new regulations consists in reducing the number of inert landfills to fewer, larger and better located sites. Most of the existing sites do not comply with the regulations, and will stop operating in the near future. Therefore, there is a need for developing and testing methods to support inert landfill site selection.

The study area is located in the south-western part of Trentino and expands within six municipalities, namely, Riva del Garda, Arco, Nago-Torbole, Tenno, Dro and Drena (Figure 3.1). The area is known as the Sarca Plain, because its morphology has been shaped by the action of the Sarca River, before entering the Garda Lake. Garda is the largest Italian lake, and one of the main tourist attractions in northern Italy. Land use is very intense due to residential expansion, industrial and tourism development, transportation infrastructures and agriculture (Figure 3.1). The area has a population of about 32,000, whose trend has shown a yearly increase by almost 1 per cent in the past decade. Over the years, urban growth has occurred around historical urban centres and along main road axes, but also in a more haphazard way. Housing and small-industry development has been encroaching into agricultural land, generating sprawl patterns, as commonly found in flat areas of north-eastern Italy. Agriculture practice is very intense, and virtually entirely devoted to vineyards. Landscape is a remarkable asset of the Sarca Plain, thanks to ruins, well-preserved historical centres, terraced hill slopes and dramatic views over the lake. The area receives over 600,000 visitors per year, attracted also by world-renowned mountain activities. There are several camping sites and hotels, mainly located close to the lakeshore. Most pressing environmental issues include land take (driven by new developments), and air pollution caused by traffic and industrial activity. The morphology induces air stagnation within the Sacra Plain, causing the area to be among the regional hotspots for air pollution.

3.2 Methods

This case study presents the application of a method based on the combination of stakeholder analysis and spatial MCA (SMCA) to first design possible sites for a landfill, and then rank them according to their suitability.

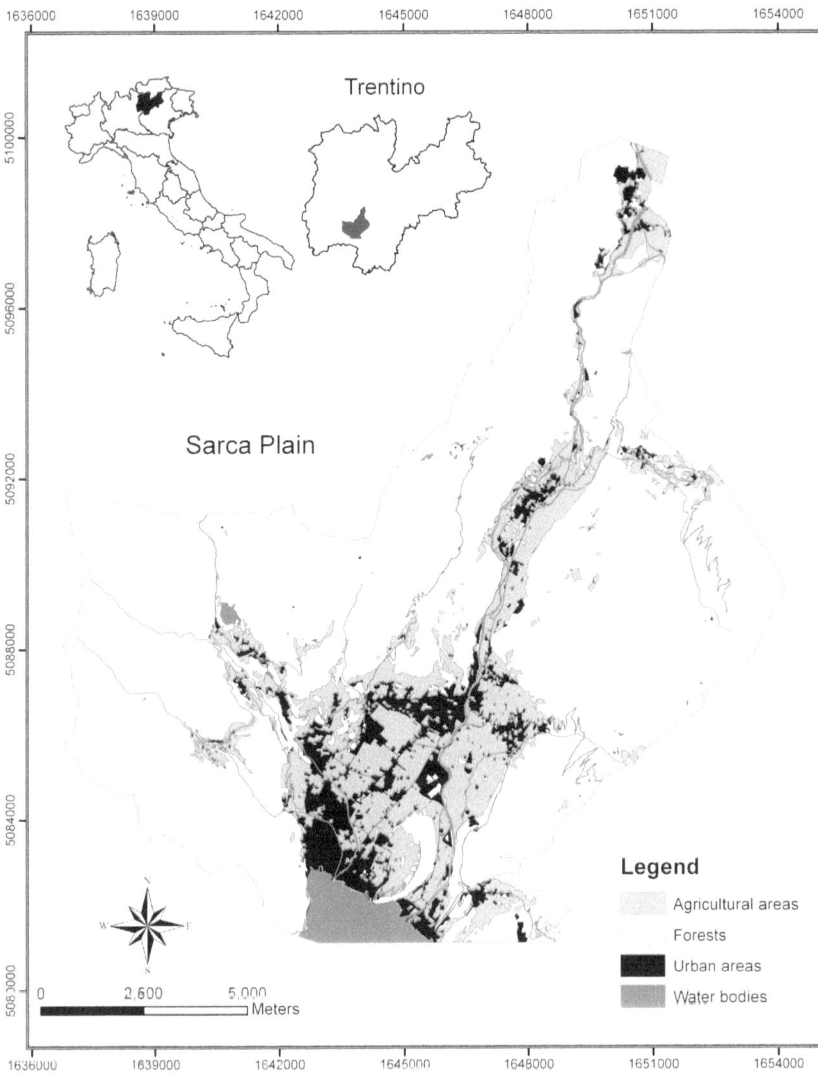

Figure 3.1 Location of the Sarca Plain (in Italy and the Trentino region) and map of the main land uses.

3.2.1 Stakeholder analysis to identify decision criteria

Stakeholder analysis is an approach for understanding a system by identifying the key actors, and assessing their respective interest in that system (Grimble and Wellard 1997). In this case study, a stakeholder analysis was carried out to identify criteria to be met by new inert landfill sites. First of all, a preliminary list of stakeholders was drawn, including environmental management offices of the Provincial Government (the Provincial Government of the Autonomous Province of Trento holds powers in environmental matters), landfill management companies, landfill users (e.g. excavation work and quarrying companies), environmental experts (universities, research institutes, etc.), environmental associations, and municipal and inter-municipal administrations. Key representatives of these groups were identified and contacted in order to get suggestions on further stakeholders to involve in the analysis. The process continued until no further stakeholders were suggested. This approach to the definition of a comprehensive list of stakeholders is commonly adopted in exercises to gather expert opinions, such as Delphi surveys (Hess and King 2002; Oliver 2002). Stakeholder groups were then classified using the approach proposed by the European Commission (EC 2003), according to which three different degrees of stakeholder involvement can be identified:

- co-operating/co-working: stakeholders actually participating in and contributing actively to the process (i.e. active involvement);
- co-thinking: stakeholders who provide input (technical knowledge and/or useful information) with respect to content and are sources of expert knowledge (i.e. consultation);
- co-knowing: stakeholders who do not play an active role in the process but should be kept informed about its progress (i.e. information supply).

This classification is useful to understand the different angles from which the problem can be viewed, and to reduce the risk of overlooking important actors.

Representatives of each stakeholder group were identified and their availability verified. The process then started by conducting face-to-face individual interviews. Individual interviews are easy to organize and do not suffer from problems commonly associated with group-based analysis, such as the fact that minority views may not receive adequate consideration due to lack of time, influence of dominant personalities and so on (Richey et al. 1985). Each participant was asked to provide a list of criteria, which were grouped into constraints and factors. Constraints are non-compensatory variables that determine areas to be excluded from the analysis. Factors are variables that contribute to the overall suitability of a site in a compensatory fashion: a poor performance of a factor can be compensated by a good performance of other factors (Eastman et al. 1995).

Finally, a thematic map was constructed for each of the identified factors and constraints. Maps were computed through basic raster GIS operations (map overlay, buffering, distance mapping, spatial queries, etc.), by using, and updating, the environmental database made available by the provincial administration (scale: 1:10,000). Data layers included land cover and land use, elevation, slope, aspect, geology, soil permeability, natural hazards, hydrology, forest inventory, ecological sensitivity and values, protected areas, road network, settlements and population. Population density information was available for census units, and was further refined by distributing population only within residential areas. Additionally, climatic data (for the past 20 years) were collected from three meteorological stations located in the northern, southern and south-eastern part of the study area, respectively.

3.2.2 SMCA to design possible landfill sites

The identification of alternative potential landfill sites was carried out in a GIS. The preliminary step consisted in applying the spatial constraints identified by the stakeholders (e.g. protected areas, areas too close to settlements). To this purpose, the constraint maps were merged and used to mask unsuitable areas, which were excluded from the analysis. Within the remaining areas only, maps representing the factors identified by the stakeholders were combined by following the typical steps of MCA (see Figure 1.1). Factor assessment was performed by using linear functions that converted the original scores (each expressed in its own unit of measurement) into dimensionless scores in the 0–1 range. Linear functions were preferred for the sake of simplicity, and were drawn by identifying 'endpoints', that is, values that correspond to the minimum (0) and maximum (1) suitability level (Beinat 1997). For example, say that among the factors there is terrain slope, and that the endpoints are set to 5° and 20°. This means that areas steeper than 20° are assigned a suitability value of 0, and areas with slope values smaller than 5° are assigned a suitability value of 1. All other areas receive intermediate values that result from linear interpolation between the two endpoints. Weights were assigned to each factor, according to its relative importance. A four-level linguistic scale (very important, important, moderately important and scarcely important) was used for this purpose.

In order to select the most appropriate endpoints and weights, a second round of stakeholder interviews was conducted. Stakeholders were first introduced to the meaning and implication of the assessment and weighting procedure, and then asked to provide, through individual questionnaire sheets, endpoints and weights for all factors previously identified. The results were processed by computing, for each factor, the average value of endpoints and weights. The latter was obtained by transforming the linguistic scale into a numerical one, and then adjusting the weight values so that each weight ranged from 0 to 1, and the sum of all weights was 1 (Malczewski 1999). The obtained endpoint allowed to construct the linear functions that were then applied to normalize the factor maps. Subsequently, the normalized maps were combined through weighted summation, according to the formula (Eastman et al. 1998):

$$S = \sum_{i=1}^{n} w_i x_i , \quad (1)$$

where:

S: suitability score;
n: number of factors;
w_i: weight assigned to the factor i; and
x_i: normalized score of factor i.

As a result, a suitability map was obtained, in which each cell was assigned a score in the 0–1 range expressing its degree of suitability to host an inert landfill. The map was then processed in order to extract most suitable cells. To this purpose, all cells with a suitability score larger than 0.80 were first selected. Subsequently, thresholds based on size and compactness were applied in order to filter out areas smaller than two hectares (the minimum area requirement for constructing and operating a landfill), and areas too irregularly shaped. This resulted in the identification of potential landfill sites, which were subsequently compared, as described in the next subsection.

3.2.3 MCA to rank landfill sites

A second set of factors was used to compare and rank the performance of the previously identified potential landfill sites. Factors included in this set differ from those previously employed because they can be meaningfully assessed only with reference to already identified sites, rather than generally for the whole study area. Factors, as well as their weights, were identified through interviews with five researchers with a publication record in relevant topics. This approach was justified by the more technical nature of this part of the problem. The proposed factors were: visibility of the site, dust pollution and accessibility. Site visibility was assessed through viewshed analysis in a GIS (Burrough and McDonnell 1998), which generated a map of the portions of landscape visible from each potential site. The results were crossed with a map representing population density in order to compute the number of inhabitants visually affected by the presence of the landfill. Dust pollution was computed through a GIS model, in which topography and wind direction were employed to estimate the distribution of dust around each site (EPA 2007; EC 2006). Also in this case, the results were overlapped with population density to estimate the number of people potentially affected. Finally, accessibility was assessed by computing the length of the new road link required to connect each site with the closest major road. As a result, an evaluation matrix was constructed, containing the performance scores of each site with respect to each of the three factors.

The evaluation matrix was used as input to a new multicriteria evaluation aimed at ranking the potential landfill sites. Factor scores were normalized using a 'maximum standardization' function (Janssen et al. 2001):

$$\text{Normalized score} = 1 - (\text{Actual score}/\text{Maximum score}). \quad (2)$$

This approach was preferred to the 'endpoint approach' adopted in the previous stage because no other information was available to select critical thresholds. Moreover, it offered the advantage of maintaining the ratio between the normalized and the original scores. The normalized scores were then combined through weighted summation (1) in order to rank the sites, by using the weights suggested by the interviewees.

Finally, sensitivity analyses were applied to assess the stability of the obtained site ranking, with respect to variations in factor scores and weights. First, perturbations were imposed on the scores to verify their influence on the overall performance of the sites. An uncertainty range was identified, representing the maximum percentage by which we expect the score to deviate from its original value, due to errors in the models or data inaccuracy. A procedure that generates 10,000 random numbers for each score (uniformly distributed within the predefined uncertainty ranges) was adopted. The results were used to run 10,000 multicriteria evaluation. They were summarized in a frequency table that shows how often each site ranked in each position. Finally, the sensitivity of the ranking was tested with respect to changes in the factor weights. The test was performed by computing rankings for all possible values of a selected weight. During this procedure the ratios of the remaining weights remained the same as those of the original weights. This analysis emphasized the reversal points, that is, the weight values that cause a reversal in the rank order of the sites. Hence, it showed the sensitivity of the results with respect to deviations from the original weight set. All sensitivity analyses were performed by using the software DEFINITE 2.0 (Janssen et al. 2001).

3.3 Results

The results of the stakeholder analysis showed that most stakeholders fell in the co-thinking group: although they do not actively participate in the process, they are considered as a source of knowledge and can provide expert judgement. This group included technical officers of the provincial government

(Geological survey, Wildlife and forest agency, Landscape protection agency), people living in the area, universities and research centres, environmental associations, as well as private bodies, such as mining and excavation work companies, farmers' associations and environmental consultancies. Fewer stakeholders belong to the co-operating group: they are the people actually involved in selecting the site, granting the authorization and managing the landfill. Interviews were conducted with representatives of eight stakeholder groups, which included different fields of expertise and perspectives.

The results of the interviews consisted of a set of constraints and a set of factors, the latter ones provided with indications on endpoints and weights. However, in two cases the interviewees did not indicate endpoints. As expected, the constraints that emerged from the interviews comprised all the requirements that were listed in the new regional guidelines, but also additional elements. More in detail, constraints concerned water (river and lake buffer zones, spring and well protection areas), nature (wetlands, forests, parks and areas designated for conservation in general), terrain (steep and mountainous areas), human assets (settlements and buildings, roads, green and recreation areas, sites of cultural and historical value) and natural risks (areas prone to floods and landslides). The set of factors included: distance from settlements, elevation, slope, distance from water bodies, presence of prime farmland, soil permeability and ecological value. The first four factors are represented by continuous variables, which were normalized into a 0–1 range using the endpoints provided through the interviews (see Table 3.1). The prime farmland criterion is represented by a Boolean map, which was normalized by assigning 1 to the absence of prime farmlands, and 0 to their presence. Soil permeability and ecological value were assessed by resorting to existing categorical maps. The permeability map, originally classified in five classes (from very high to very low), was normalized by assigning 1 to the 'very low' class, 0.75 to the 'low' class and so on. The same process was applied to the ecological value map, which was originally classified into five levels of ecological significance (details can be found in Geneletti (2008a)). The average weights assigned by the stakeholders to the six factors are presented in Table 3.1.

The weighted combination of factors generated the suitability map presented in Figure 3.2. A large portion of the study area (almost 90 per cent of it) was considered unsuitable for effect of one or more constraints. This was largely expected, due to the intensity of current land uses in the area, the complex morphology and the high number of features that acted as constraints. Suitable areas are mainly scattered within the margins of the valley floor, often close to industrial sites and transportation infrastructures. After applying thresholds based on size and compactness, eight possible sites were identified (see Figure 3.3). Sites 1, 2 and 8 share very similar characteristics: they are

Table 3.1 Factors used to construct the land suitability map, with indications on their endpoints (the first endpoint corresponds to the lowest suitability value, the second one to the highest) and relative weight

Factor (unit of measurement)	First endpoint	Second endpoint	Weight
Distance from settlements (m)	200	1000	0,21
Elevation (m)	1200	100	0,08
Slope (°)	0	20	0,13
Distance from water bodies (m)	600	150	0,15
Soil permeability (qualitative classes)	Very high	Very low	0,07
Prime farmland (Boolean)	Present	Absent	0,19
Ecological value (qualitative classes)	Very high	Very low	0,17

found within a 1-km radius in the narrower part of the Plain. The current land use is agriculture and the sites are located in the outskirts of a small village, close to the Plain's margins. Site 6 is located at an elevation of about 500 m on a flat plot of vacant land, surrounded by agriculture and woodlands. Site 7 is adjacent to an industrial area in Riva del Garda, the largest municipality of the study region. Current land use is small-scale agriculture. Sites 3 and 4 are located on the eastern margin of the Plain in vineyard-covered areas. Finally, Site 5 is found on a plot of vacant land close to a quarry, surrounded by woodlands and shrubs. The average suitability value of the eight sites is very similar, ranging from 0.72 (Site 7) to 0.79 (Site 3). Therefore, it would have been very hard to select the most suitable site using this information alone.

The second set of factors (visibility of the site, dust pollution and accessibility) was computed for the selected sites. The results are presented in Table 3.2, together with the weights assigned to each factor. Unlike the average suitability computed earlier on, the scores of these factors present a wide range of variations among the sites. Hence, the use of the additional set of factors helped to clearly differentiate the performances of the sites. Four sites do not present any impact in terms of dust pollution (Sites 2, 3, 5 and 8), whereas the remaining four are expected to affect from few tens to few hundreds of the residents. According to the viewshed analysis, Site 5 is the less disturbing, whereas Sites 4, 7 and 6 are the most visible ones, being located close to the most densely populated areas. As to accessibility, none of the sites is remotely located, and the distance from a major road ranges from 110 to 680 meters. Sites 4 and 6 are the most favourably located in terms of transportation infrastructures, whereas Sites 3 and 5 are the most disadvantaged. The weighted summation of the normalized factor scores generated the results presented in Figure 3.4. This figure shows also the relative contribution of each factor to the overall

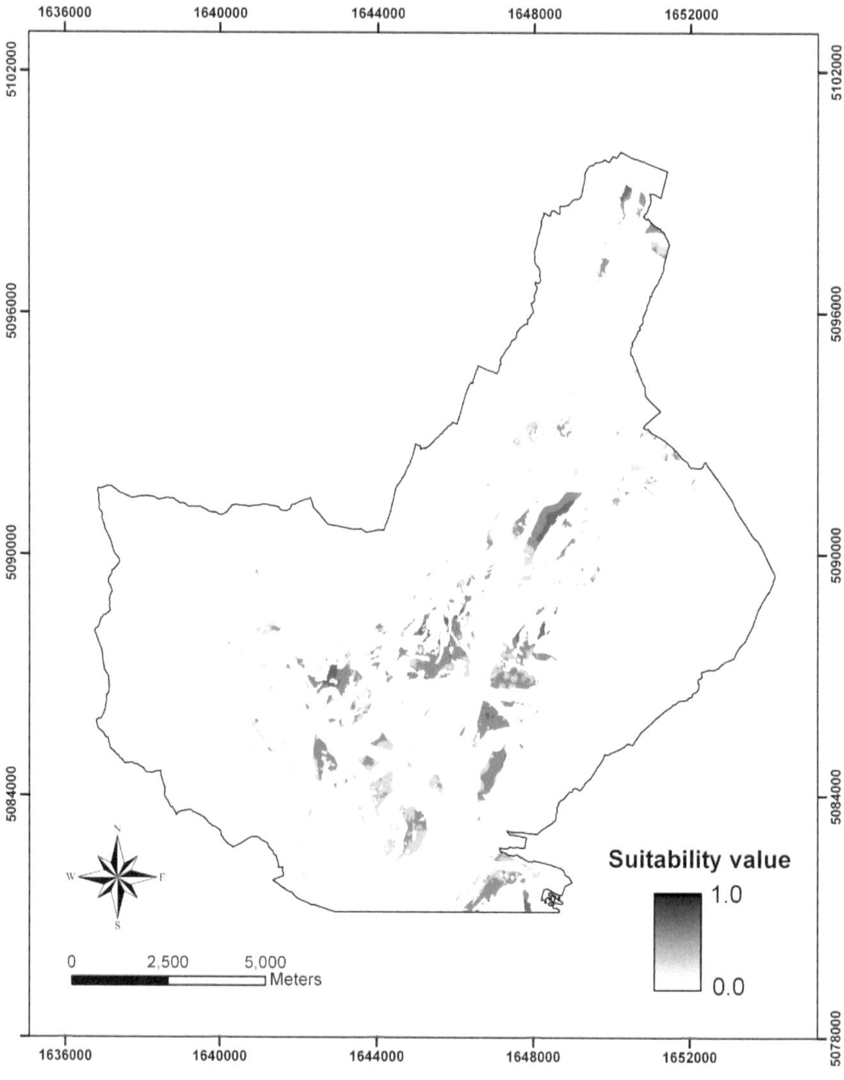

Figure 3.2 Land suitability of the Sarca Plain to host an inert landfill (1: highest suitability, 0: lowest suitability).

performance score of the sites. Sites 8, 2 and 1 ranked at the top, Sites 6, 5 and 3 in the middle, and sites 4 and 7 at the bottom. The clustering of the sites is evident also in geographical terms: the best three sites are located very closely in the northern sector of the valley (Figure 3.3).

The analysis of the sensitivity to changes in the factor scores was performed by considering a ±30 per cent uncertainty range for visibility and dust, and a

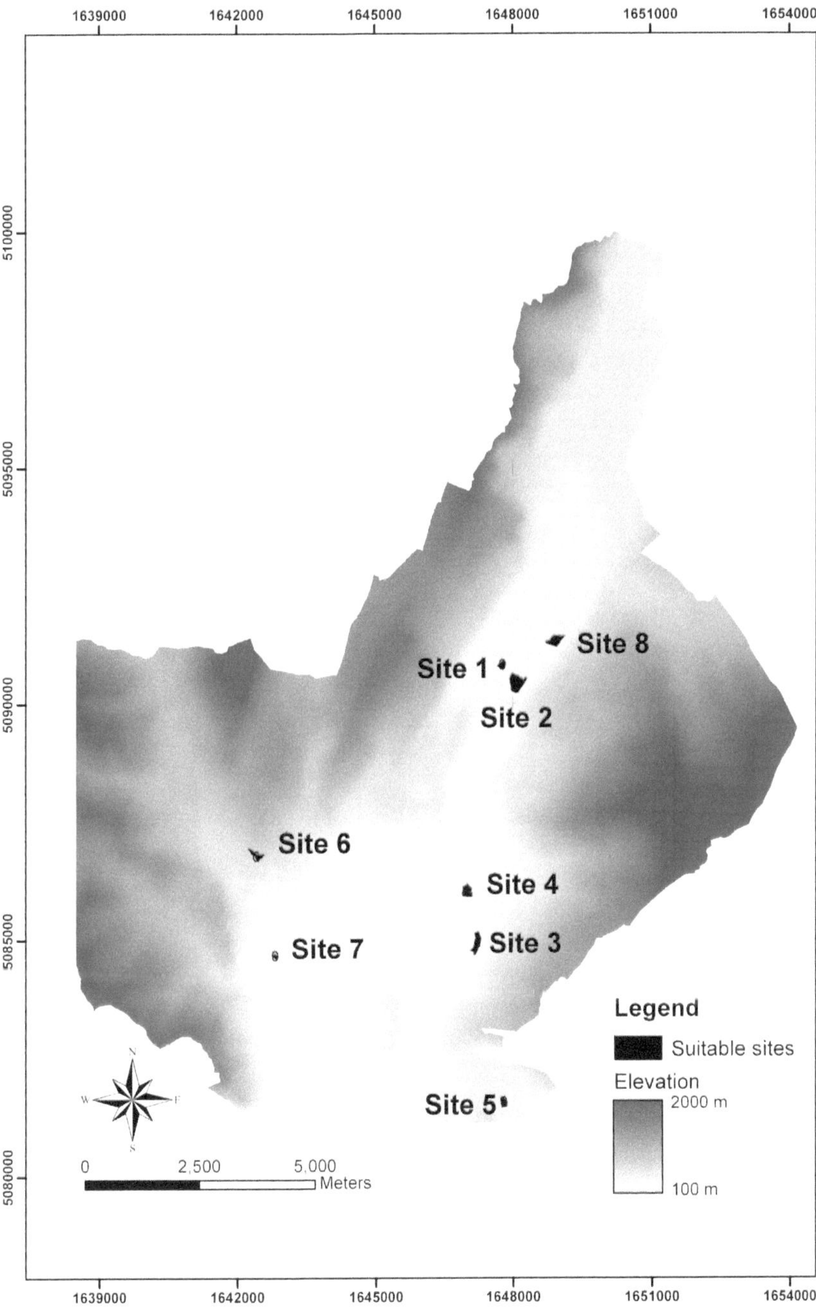

Figure 3.3 Location of the eight potential landfill sites superimposed on to the elevation map of the area.

Table 3.2 Evaluation matrix showing the performance score of the potential landfill sites with respect to the three factors considered for site comparison

Factor (unit of measurement)	Weight	Site 1	Site 2	Site 3	Site 4	Site 5	Site 6	Site 7	Site 8
Dust (no. of people)	0.397	33	0	0	250	0	81	276	0
Visibility (no. of people)	0.317	69	97	564	5874	23	1002	7805	79
Accessibility (m)	0.286	250	245	685	110	650	140	290	190

Note: The relative weight of the factors is also provided.

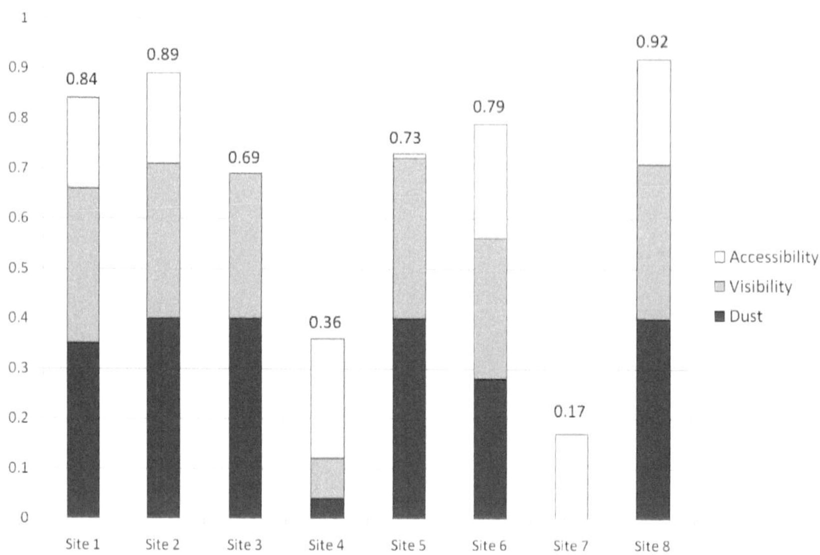

Figure 3.4 Ranking of the sites with indication of the relative contribution of factors considered (visibility of the site, dust pollution and accessibility).

±10 per cent range for accessibility. The rationale was that the first two factors required more complex modelling, and therefore were potentially affected by larger errors. The frequency table summarizes the ranking obtained in the 10,000 iterations (Table 3.3). Results are presented only for Sites 8, 2 and 1 because the remaining ones always ranked lower, and were therefore irrelevant. The frequency table confirmed the dominance of Site 8, which ranked first 51 per cent of the times. Sites 2 and 1 ranked in the first position only 25 per cent of the times. Another interesting result is that the fluctuation range is wide for Site 1 (from the first to the sixth position), and much narrower for sites

Table 3.3 Frequency table resulting from the 10,000 simulations that were run after introducing uncertainty ranges in the factor scores

	1st	2nd	3rd	4th	5th	6th
Site 8	51	41	8	0	0	0
Site 2	25	44	29	2	0	0
Site 1	25	14	30	13	8	11

Note: The table shows how often (in percentage) each site ranked in each position. Results are presented only for the best-performing sites.

8 and 2 (first three positions, except 2 per cent of the times for Site 2). This means that errors in the scores have a higher effect on Site 1, whereas Sites 8 and 2 proved to be more stable.

Finally, sensitivity analysis was applied to weights, by computing the ranking for all possible values of the weight of a factor at a time. The results are summarized in the graphs of Figure 3.5. As can be seen, for all the three factor weights, there are no reversal points between Sites 8, 1 and 2 (the three lines that represent these sites never crossed one another). That means that Site 8 ranks first, Site 2 second and Site 1 third, no matter how much the value of a weight is changed. Some rank reversals did occur with Sites 6 and 4, but the reversal points are located very far from the original weight value. Hence, the sensitivity analysis showed that the top part of the ranking is stable, and that Site 8 is to be preferred under most conditions. It is also crucial to remember that the three top-ranking sites are located close to each other. Therefore, another clear result of the study is that that specific portion of the study area is by far the most suitable to host an inert landfill, in the light of the siting criteria that were considered.

3.4 Lesson Learned and Conclusions

Spatial MCA is commonly applied to land suitability analysis (Malczewski 2004), and specifically to landfill site selection studies (Şener et al. 2006; Mourmouris 2006). However, relatively few studies incorporate the opinion of stakeholders and the public (Higgs 2006): in most applications, the value-based input (e.g. weights of the different criteria; critical thresholds of the adopted indicators) is provided by the same authors or by a panel of technical experts, through techniques such as interviews or Delphi surveys. This generates results that might be technically appropriate, but not necessarily suitable in terms of adjustment to the local socioeconomic context. On the contrary, a decision process able to account for the full range of

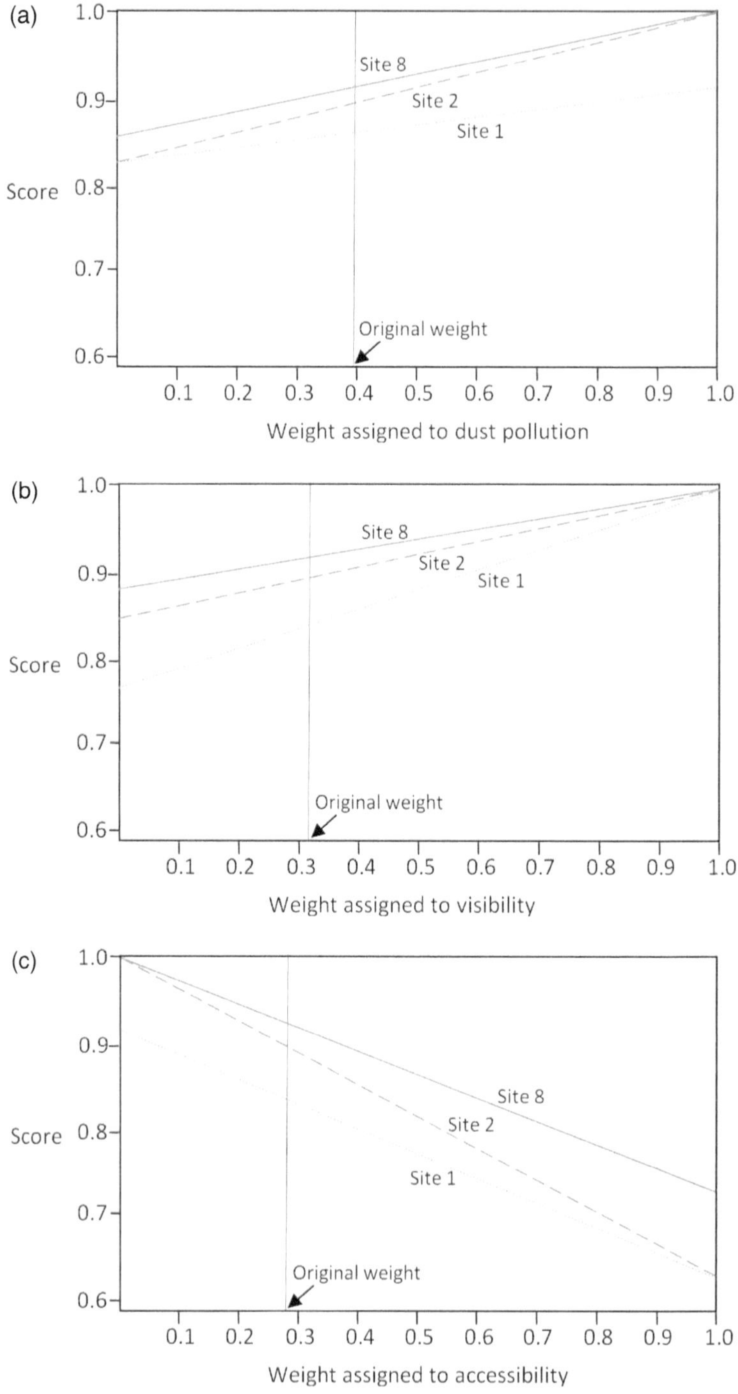

Figure 3.5 Sensitivity of the site ranking with respect to changes in the weight assigned to (a) dust pollution, (b) visibility of the site and (c) accessibility.

values and concerns can be more successful in finding acceptable solutions (Petts 2001).

This case study illustrated an approach involving stakeholders' opinion that can be replicated by the local administration to support landfill siting processes. The two-stage approach allowed to first identify potential inert landfill sites within the study region, and then rank them according to their preferability. The two stages were conducted using different sets of criteria, and inputs from different groups of people. In the first stage, general factors and constraints were suggested by a broad group of stakeholders, and applied to the whole area to generate a suitability map. The most suitable areas were then extracted by using also thresholds based on size and shape. This was required in order to select reasonably large and compact sites. In the second stage, criteria that could be assessed specifically for predefined sites were suggested by experts in the field. This allowed accounting for some important factors, such as the number of people potentially affected by dust pollution.

The two-stage approach proved to be effective in better differentiating the performance of the sites, and thus in identifying the most preferable one(s). The involvement of different groups of people (i.e. stakeholders and experts) during the two stages allowed handling the different types of inputs required given the increasing technical nature of the problem. Stakeholder analysis and interviews proved to be time-saving tools to gather people's perception and values. This is particularly important for the location of small landfills, since a fast decision-making process is often required. The results were strengthened by sensitivity analyses. Sensitivity analysis is an important tool that can be applied to account for errors and inaccuracies, as well as for different perspectives and opinions, as shown in this research. As such, it is useful to support planning processes that typically face public opposition, such as the location of landfills.

Chapter 4

PROTECTED AREA ZONING: INTEGRATING SPATIAL MCA AND MULTI-OBJECTIVE EVALUATION*

4.1 Context

4.1.1 Environmental decision-making problem

Optimizing the multiple use of land is a challenge for today's governments and land managers. To achieve this, they need to assess land properties for their environmental, economic and social importance and derive suitable legislation and management plans. These plans should satisfy the demand for multiple land uses, while protecting natural resources. In protected area planning, land properties are assessed with different, possibly conflicting, objectives. Production, hunting, fishing and recreational use cause disturbances, such as direct reduction of species populations, habitat degradation and fragmentation. These disturbances in turn may influence ecosystem composition and processes (Allison 1996; Hobbs 1993; Moleele and Mainah 2003; Noss 2000) or change behavioural patterns of species (Creachbaum et al. 1998).

Park planners and managers consequently face the problem of selecting the most appropriate evaluation framework and indicators. Planning and management decisions will also be strongly influenced by the type and amount of information already available, the size of the area and the legislation within which the planning exercise needs to fit. Arguably, one of the most important elements of a protected area plan is the zoning scheme. Through zoning, specific uses – and constraints – are assigned to different land units. Typically, zoning schemes consist of core areas, where strict nature conservation is enforced, and areas where gradually more intensive human presence and activities are allowed. In practice, planners and managers need to evaluate the

* The text and graphics of this chapter are based on the following paper: D. Geneletti and I. van Duren, 'Protected Area Zoning for Conservation and Use: A Combination of Spatial Multicriteria and Multiobjective Evaluation', *Landscape and Urban Planning* 85 (2) (2008): 97–110.

spatial distribution of land properties, and decide upon where to restrict or stimulate certain activities or where to implement measures to protect natural resources.

Setting the zoning scheme for a protected area is a decision-making problem that inherently requires the evaluation of multiple land attributes according to multiple objectives. In addition, the processes that lead to a zoning proposal should be transparent, and the evaluation of land properties scientifically sound. Many studies have revealed the potential contribution of MCA to similar processes (Bojórquez-Tapia et al. 2004; Crossman et al. 2005; Hjortsø et al. 2006; Keisler and Sundell 1997; Ridgley and Heil 1998; Villa et al. 2002).

4.1.2 Case study description

This case study concerns the zoning of a terrestrial protected area: the Paneveggio-Pale di S. Martino (PPSM) Natural Park, located in north-eastern Italy. The Park represents an example of the need to combine several potentially conflicting objectives, such as nature conservation, water resource protection, tourism and recreation, cattle grazing, preservation of outstanding geomorphologic sites, traditional landscapes and scenic views. This study was initiated to support the drawing of the new spatial plan for the PPSM Natural Park (eventually approved in 2016), and the associated new zoning scheme. According to national and regional regulations on protected areas planning, zoning needs to include three main protection levels:

- Zone A: strict protection of the environment and the ecosystems, minimizing presence of or disturbance by human activities;
- Zone B: protection of cultural, historical and landscape assets, restricting land use to traditional activities that are considered compatible with environment protection;
- Zone C: promotion of recreational use and development of low-impact tourist facilities, minimizing the disturbance to the environment.

The PPSM Park covers about 20,000 ha. Elevation ranges from around 1,100 m in the south-eastern valleys and the highest mountain peaks reach over 3,100 m asl. The Park is well known for its landscape beauty and its spectacular alpine geomorphology attracting many tourists for summer and winter outdoor activities (Figure 4.1). Additionally, the Park contains more than 1,500 different plant species and a large variety of animal species, among others roe deer (*Capreolus capreolus*), elk (*Cervus elaphus*), chamois (*Rupicapra rupicapra*), Alpine Marmot (*Marmota marmota*), European hare and brown hare (*Lepus*

Figure 4.1 A typical landscape in the Paneveggio-Pale di S. Martino Natural Park. *Source*: Picture by Davide Geneletti.

europaeus and *Lepus timidus*). Among the avian fauna, interesting species, such as rock partridge (*Alectoris graeca*), hazel grouse (*Tetrastes bonasia*), capercaillie (*Tetrao urogallus*), Eurasian black grouse (*Lyrurus tetrix*) and the rock ptarmigan (*Lagopus mutus*) are found.

In Zone A of the Park only scientific research, educational excursions, hiking on paths and maintenance of already existing facilities are permitted. In Zone B, land use is restricted to types that are considered compatible with environmental conservation. Here, regulated grazing, walking off the regular hiking paths, as well as some traditional agricultural activities are permitted. Forest maintenance aims at safeguarding forest functions such as protecting the hydrological resources and contributing to landscape dynamics. Natural forest developments and regrowth of native species are stimulated, though active reforestation is not practiced. However, the wild populations of large herbivores in the forests are controlled to limit damage by overgrazing and browsing. Controlled cattle grazing in pastures is allowed, but the use of pesticides and herbicides is forbidden. It is also not allowed to influence the hydrological patterns. In Zone C, recreational use and the development of tourist facilities are promoted, although efforts are made to minimize the disturbance on the environment. Facilities that can be found in this zone include camp sites, hotels, picnic areas, ski pistes, ski lifts and parking places.

4.2 Methods

The Park was subdivided into land units to be used as the basic spatial elements of the zoning mosaic. Land units are homogeneous in terms of elevation, slope range, geology, vegetation cover and land use. A combination of spatial MCA and multi-objective land allocation (*sensu* Eastman et al. 1998) was then run to assign a protection level to each land unit (Figure 4.2).

4.2.1 Spatial MCA to assess land suitability

Three spatial MCA were performed to map the suitability of the Park land units to be assigned to Zones A, B and C, respectively. Figure 4.2 shows a simplified version of the criterion trees used for the land suitability analysis. In Zone A strict conservation is enforced, therefore decision variables are related only to nature conservation and include habitats, species and outstanding natural features. Habitats were assessed according to rarity and relevance to nature conservation. Rarity was measured at the local level by computing the cover percentage of similar habitat types within the Park. Relevance to nature conservation was expressed using a 1–5 scale. The top two levels were assigned, respectively, to priority and non-priority habitat types according to the European Union's Directives on nature conservation. The remaining three levels were assigned by expert botanists. As to species, both their actual and potential presence was included in the analysis. Their value was assessed by considering the presence of the species in the Italian Red Lists of threatened species and in the lists provided by European Union's Directives on nature conservation. Finally, outstanding natural features were mapped by aggregating data layers containing the distribution of natural springs, waterfalls, monumental trees (e.g. remarkable old trees), glaciers, glacial lakes and geosites. Geosites include significant geomorphologic and geologic features, such as fossil areas, moraine ridges, peculiar outcrops, as well as the well-known Pale di S. Martino limestone cliff after which the Park is named.

In order for the maps to be comparable, they were then normalized using a 0 to 1 range. Boolean maps, such as the ones depicting outstanding natural features, were normalized by assigning 1 where those features are found, and 0 elsewhere. All other maps were normalized by applying the "maximum" method, which offers the advantage of keeping the ratio between actual and normalized values. If a criterion has a positive relation with the suitability

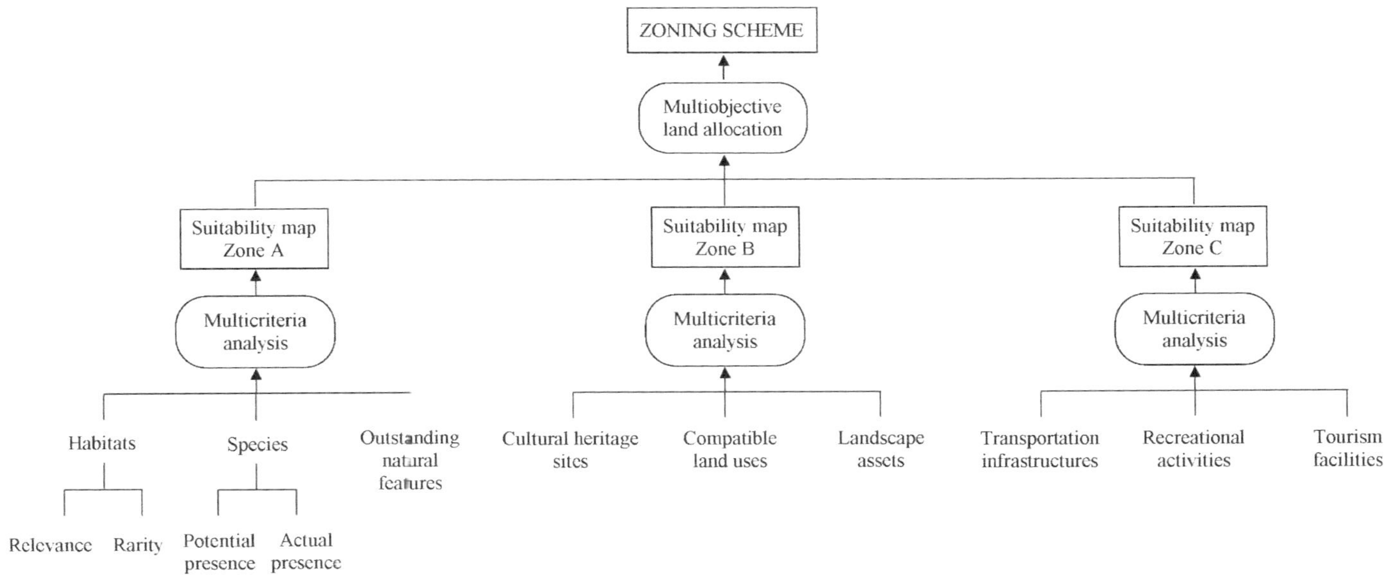

Figure 4.2 Flowchart of the method and evaluation criteria for the three protection levels (zones).

for Zone A (i.e. the higher the criterion value, the higher the suitability), the maximum normalization was applied by dividing the actual score by the maximum score. If the opposite applies (i.e. the higher the criterion value, the lower the suitability) the formula is, according to Malczewski (1999):

$$\text{normalized value} = 1 - (\text{actual score} / \text{maximum score}). \qquad (1)$$

The normalized criterion maps were combined through weighted summation. This led to the generation of a land suitability map for Zone A. The weights assigned by the authors were given after consulting with Park experts and managers. First, qualitative weights were elicited, that is, an ordinal priority ranking of the criteria was established. The ranking was then used to derive quantitative weights through direct assessment (Table 4.1). To factor in that stakeholders and experts might have different views on the relative importance of the criteria, three scenarios were run using different weight sets. These sets were selected after consultation with a group of ecologists in such a way to account for all the different viewpoints. For the sake of testing the methodology, only the weights of the criteria at the top level of the hierarchy were varied, as shown in Table 4.1. A sensitivity analysis was then performed to study the effect of different weight sets on the outcome of the land suitability map.

More simplified evaluations were carried out to support land suitability analyses for Zones B and C. Zone B focuses on preserving cultural and landscape assets, restricting land use to traditional and environmentally compatible activities. Three main criteria were considered (Figure 4.2): the distribution of cultural heritage sites, compatible land uses and landscape assets. Cultural heritage sites include archaeological areas, as well as historical and religious sites. Compatible land uses are represented by traditional land uses with positive interaction with the environment (e.g. mountain farming, which is essential for reducing soil degradation and slope instability, as well as aiding in biodiversity conservation). Landscape assets include traditional settlements and rural buildings, as well as aesthetically outstanding natural and man-made landscape features.

In Zone C recreational use of the Park and development of environmentally compatible tourist facilities should be stimulated. Three main criteria were considered: transportation facilities, recreational activities and tourist facilities. Transportation facilities include roads, parking places and public transport systems. Recreational activities include ski lifts and ski pistes, hiking trails, horse trails, picnic sites, hang-gliding and paragliding spots and so on. Tourism facilities include mountain huts, hotels, restaurants, camping sites, visitor information centres and so on.

Table 4.1 Weight sets used to generate the land suitability map for Zone A

	Weight (1)	Weight (2)	Weight (3)	Weight (1,2,3)			
Habitats	0.400	0333	0.275	Rarity	0.20		
				Relevance	0.80		
Species	0.400	0.333	0.275	Actual presence	0.80		
				Potential presence	0.20		
						Animals	0.40
						Plants	0.40
						Lichens	0.20
Outstanding features	0.200	0.333	0.450				

Like Zone A, the criteria of the decision trees for Zones B and C were normalized and then aggregated through weighted summation. Most of these criteria were represented by Boolean maps with presence/absence of the relevant features (e.g. heritage site, ski piste, etc.). In some cases, the presence was a point location or linear location (e.g. huts, trails), which was extended with a buffer area around (or along) the feature. All criteria in Zone B were considered as equally relevant. Weights for Zone C suitability were assigned by giving high importance to the presence of 'constraint' facilities, that is, facilities that should not be built or extended in the future within the protected area (i.e. road network, skiing infrastructures, etc.).

The three suitability maps were aggregated into the previously identified land unit polygons, by assigning a suitability value to each unit. Several aggregation methods can be used for this purpose. Average among all grid cells is probably the most commonly used method, but it tends to smooth out suitability values among units. The maximum value method is also often used (i.e. a unit takes the value of its best performing cell), but it has the disadvantage of highlighting units, even if they contain very small valuable features (e.g. a spring), which may not be representative of the whole unit. A compromise approach consists of taking the average value of the best performing group of cells within each unit. In this study, three methods were compared: the average value of all grid cells of the unit, the average value among the best performing 50 per cent of all grid cells of the unit and the average value among the best 30 per cent of grid cells of the unit. The 50 and 30 per cent thresholds were selected after running few tests that indicated that any other threshold value between 60 and 10 per cent would generate very similar results. The results of the three aggregation methods were then compared through sensitivity analysis.

4.2.2 Multi-objective land allocation to assign protection levels

A multi-objective land allocation procedure was undertaken to assign to each land unit the most suitable protection level. The inputs required are the three suitability maps, and a land demand for each protection level, which were after discussion with some of the stakeholders involved. In this discussion, it was suggested to adopt an allocation scheme slightly more nature-oriented compared to the current plan, where 55 per cent, 35 per cent and 10 per cent of the land is assigned to Zones A, B and C respectively. The proposed scheme features 45 per cent of the land area allocated to both Zones A and B, and 10 per cent to Zone C.

Land allocation started by selecting, for each protection level, the units with the highest suitability up to the fulfilment of the land demand. Subsequently, the results of the selection made for the three protection levels were compared. All the units that were selected for one zone only were directly allocated to that zone. Units that were selected for two or three zones, as well as units that were never selected, represented 'conflicting units'. The allocation of a particular zone in the conflicting units was performed through a decision algorithm. The algorithm optimizes choices, by taking into account simultaneously the suitability for all protection levels. This decision rule represents an adaptation to the vector environment (i.e. where units to be allocated are represented by polygons) of the raster-based approach described by Eastman et al. (1998).

The suitability maps were first reclassified by transforming the original cardinal suitability values into ordinal ones. Each land unit was given a descending rank order, by assigning one to the unit with the highest suitability value, two to the second most suitable unit and so on. This was repeated for the three suitability maps thereby making them comparable. The three suitability maps are based on entirely different criteria, determined by the different objectives. Hence, their cardinal suitability values are not directly comparable. If two units have the same suitability value for a given protection level, the resulting tied rank was solved by using as a secondary sort criteria the suitability for the other protection levels. That is, the highest rank was assigned to the unit with the lowest rank for the other protection levels.

Once the units are given ordinal ranks, the algorithm scans the conflicting units starting with the one that contains the highest rank order for any of the zones. This unit is then assigned to the corresponding protection level. If a unit has the same rank for two or more protection levels, the tie is resolved by setting priorities between the protection levels. In this exercise, Zone A received priority over Zones B and C, and Zone B received priority over C. This is consistent with the main objective of the protected area, which is to protect natural and landscape features. The algorithm iterates the procedure for the remaining conflicting units, until all of them are allocated to a protection level. At every step, a control function based on area calculation excludes further allocation to protection levels whose land demand has already been fulfilled. For example, say that, among all conflicting units, the highest rank order belongs to Unit X, and corresponds to the rank order for Zone A. That means that, among all conflicting units, Unit X is the most suitable for Zone A, and also that Unit X is relatively more suitable for Zone A than for Zone B or C. Therefore, the conflict is best solved by assigning Unit X to Zone A. The algorithm then sums the areas of the land units previously allocated to Zone A with the area of Unit X to check whether Zone A still requires land, or whether it has already satisfied its land demand. If the latter applies, only rank orders for Zones B and C are taken into account in resolving the remaining allocation of conflict areas.

The application of the decision rule allowed the conflicting units to be allocated, and hence to achieve a complete zoning of the Park for each of the two land demand scenarios. Since land units have different areas, the fulfilment of land demands was approximated to the closest value.

Finally, sensitivity analysis was run to test the stability of the results with respect to changes in the aggregation method that was used to convert raster suitability maps into land unit-based maps. Three sets of suitability maps were generated, by applying respectively the average, the average among best 50 per cent of cells and the average among best 30 per cent cells aggregation methods. The multi-objective land allocation was run three times using the different map sets as input. The resulting zoning schemes were then compared, in order to identify stable units, that is, units whose allocated protection level was not affected by the aggregation method, and unstable units, which on the contrary changed protection level depending on the aggregation method.

4.3 Results

The land suitability maps for Zones A, B and C are shown in Figure 4.3. Highest suitability values for Zone A are found mainly within three regions: the highest peaks in the eastern sector of the Park, the north-western slope of the mountain range that crosses the central sector of the Park and the southern slope of the northernmost peaks. The first region stretches at elevation ranging from 2,500 to 2,900 m, and it is covered by bare rocks. The second and third regions are located at around 2,000 m, and their predominant land cover is represented by natural grasslands and coniferous forest. Most suitable areas for Zone B are small in size and scattered in the central and south-western sector of the Park. They correspond to areas where traditional land uses or remarkable landscapes overlap with sites of cultural or historical

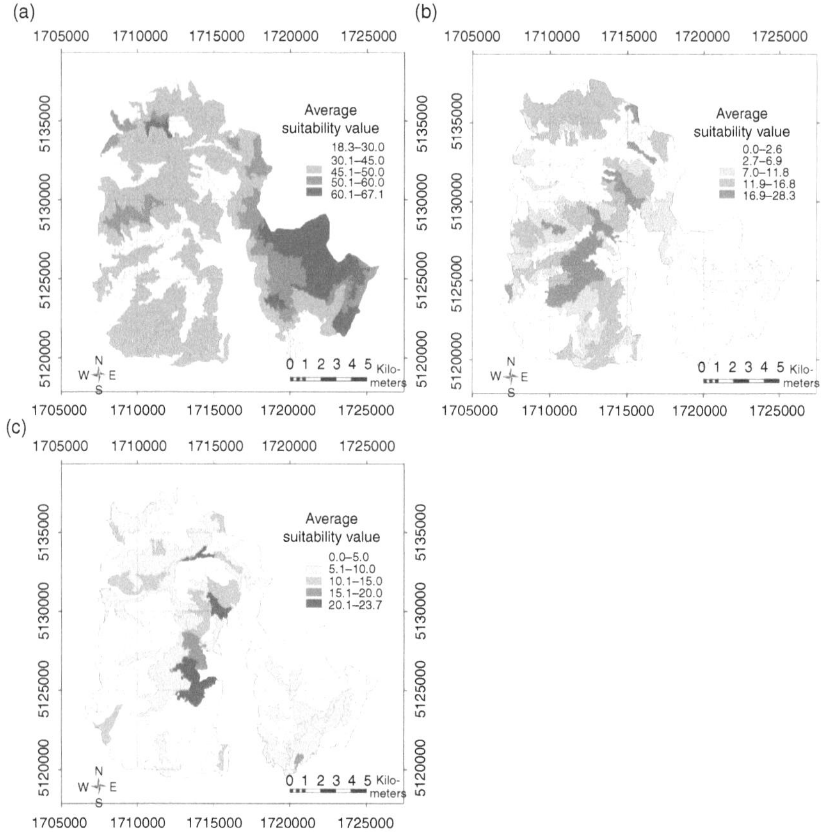

Figure 4.3 Land unit suitability maps for Zones A (a), B (b) and C (c).

interest, which are limited in size. However, there are also large and connected expanses of medium to highly suitable areas in the central-western sector of the Park. Even though the suitability values for Zone C are mostly distributed according to linear and point elements (roads, trails, huts, etc.), two highly suitable regions stand out. Both of them are located in the central sector, along the Park boundaries, and host major ski centres. The northernmost area is developed and heavily affected by human disturbances, due to the construction of hotels and other tourism facilities. The other area is better preserved, and includes only a few huts.

Sensitivity analyses were run to test the stability of the results with respect to altered weights and differing aggregation methods. The sensitivity to changes in weights was conducted only for Zone A, and aimed at assessing the robustness of the relevant suitability map. The effects of applying three different sets of weights (Table 4.1) are presented in Figure 4.4. These figures compare the rank order of each land unit for each pair of weight sets. Dots that lie on the line represent units that rank in the same position using both weight set. Only a few of the units rank in a remarkably different order. These units are primarily found in the middle range of the ranking, whereas the most and least suitable units are less sensitive. This is due to the fact that in the best and worst units there are cells characterized by respectively favourable and unfavourable conditions for all criteria. These cells are less sensitive to changes in weights, and they influence the overall performance of the unit, which therefore tends to be stable. On the contrary, units in the mid-sector of the ranking have fewer cells with extreme values, and their performance is influenced by cells which present favourable conditions for some of the criteria, and less favourable for other ones. Such cells are more sensitive to weight changes, hence they influence the ranking of the unit to which they belong. In light of these results, it was decided to carry out land allocation using only one suitability map. The map obtained with Weight set 1 was chosen because this set was preferred by the group of people involved in the assessment.

The sensitivity analysis with respect to aggregation method was carried out by applying the three methods described in Section 4.2, and then comparing the resulting maps. The first column of Table 4.2 shows the average change in the suitability values for the three zones. It was computed as the average of the percentage of change between each pair of maps. The second column of Table 4.2 shows the corresponding average shift in the unit position upward or downward in the ranking. It was computed after transforming the cardinal values into rank orders. Zone C is the most affected in terms of change in the cardinal suitability values. However, these differences have a limited effect on the relative performance of the units, which shift on average about three positions in the ranking. On the contrary, the ranking for Zone A is significantly affected by the aggregation method, with an average shift of over ten positions.

The zoning schemes resulting from executing the land allocation algorithm obtained with the three aggregation methods are presented in Figure 4.5 (a, b, c), and compared in Figure 4.5 (d). Stable units cover over 70 per cent of the Park. The eastern sector of the Park was assigned to Zone A. This area is characterized by outstanding geomorphologic features, such as dolomite peaks, screes and glaciers. Additionally, it is dotted by priority habitat types (e.g. bushes with *Pinus mugo* and *Rhododendron hirsutum*, limestone pavements). Most of the south-western sector of the Park was allocated to Zone B in all the three zoning schemes. Traditional land uses, cultural heritage sites and open landscapes are characteristic features of this area. Predominant land covers are temperate mountainous coniferous forests and Alpine and Boreal heaths

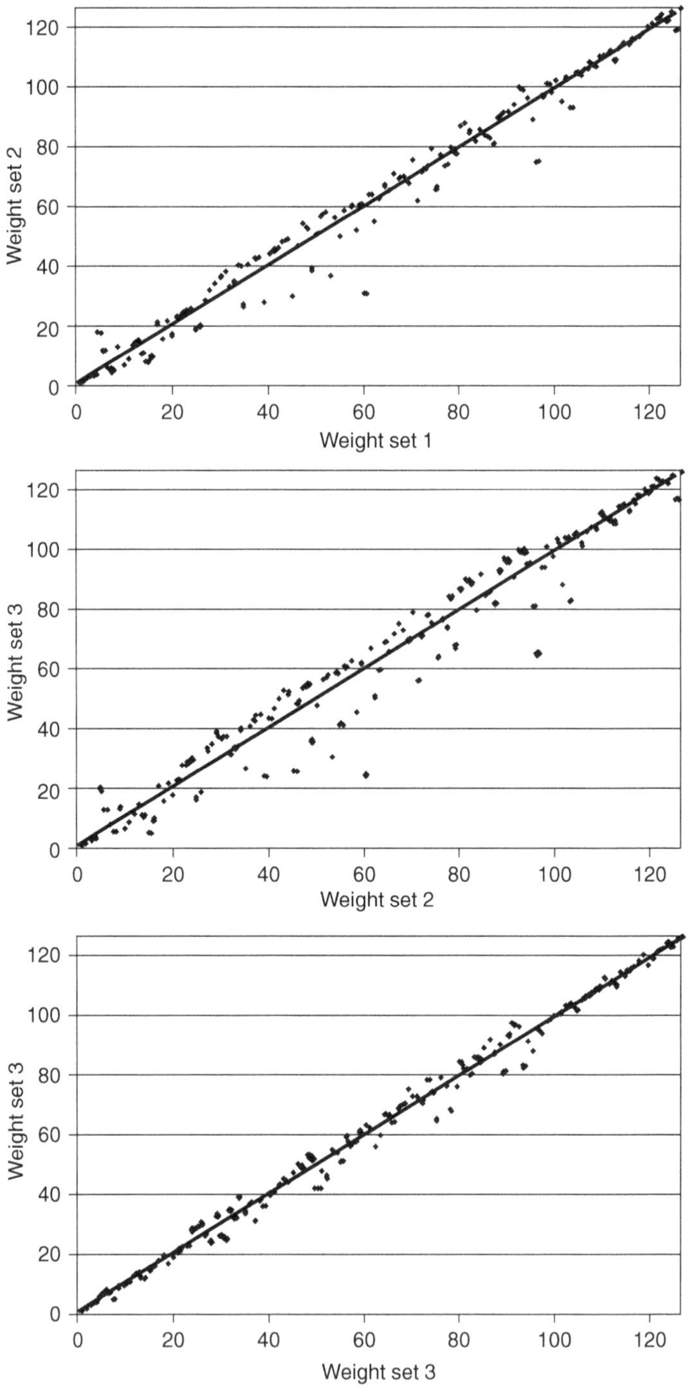

Figure 4.4 Comparison of land unit rankings obtained with the three weight sets (Zone A).

Table 4.2 Results of the sensitivity analysis with respect to aggregation methods

Zone	Average change in suitability value (per cent)	Average shift in rank order
A	7.3	10.7
B	18.6	3.6
C	29.8	4.4

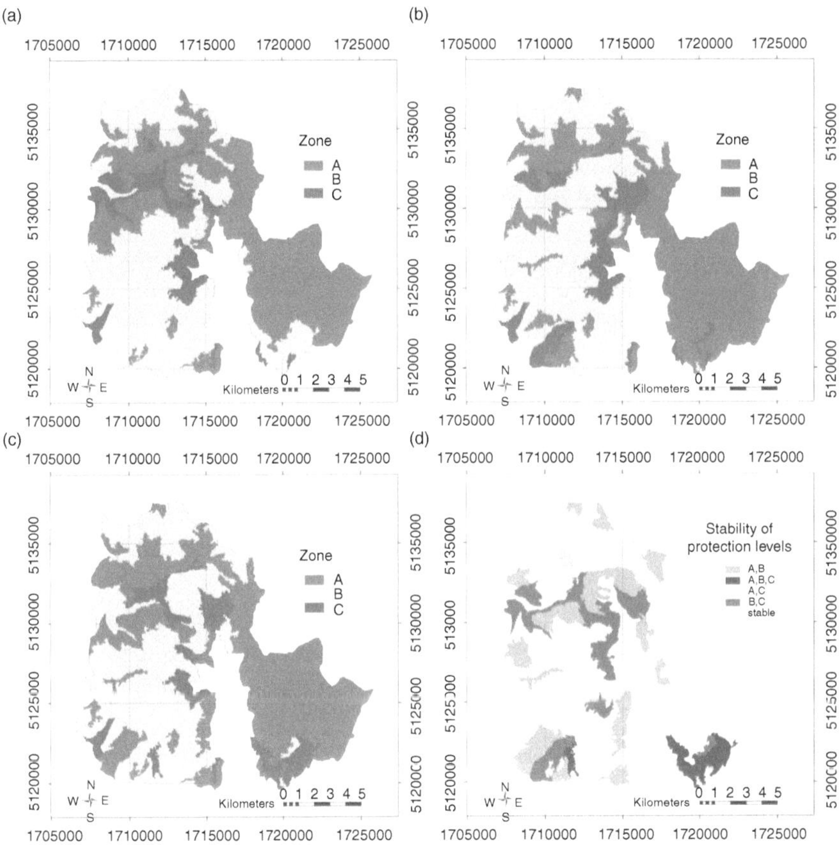

Figure 4.5 Zoning schemes obtained with the three aggregation methods (a: average; b: average among best 50 per cent cells; c: average among best 30 per cent cells) and comparison (d).

and grasslands. Units assigned to Zone C are located in the central part of the Park, due to the presence of ski infrastructures and other tourism facilities.

Units that fluctuate between Zones A and B cover about 10 per cent of the Park area. They are scattered along the Park boundaries, as well as in the central region. In these units, landscape assets and compatible land uses coexist with rare habitat types and animal species. Units that fluctuate between Zones B and C cover almost 5 per cent of the area, and are mostly found in the central sector of the Park. These units are characterized by the presence of features of interest for Zones B and C that are limited in size (e.g. cultural heritage sites, huts, trails). Therefore, they are very sensitive to the aggregation method, which can highlight or smoothen out the overall suitability value of a unit. Finally, two very small land units fluctuate between Zones A and C, and other two land units (covering about 3 per cent of the area) fluctuate between the three protection levels. The latter ones are located along the southern boundaries of the Park, and are characterized by anthropogenic elements (camping site, roads, etc.), but also by traditional mountain land uses and small patches of priority habitats (species-rich Nardus grasslands on silicious substrates).

4.4 Lesson Learned and Conclusions

The approach allowed park managers and other stakeholders to visualize and understand the processes that lead to the zoning scheme in a clear and transparent way. Furthermore, it gave the opportunity to thoroughly test the methodology and evaluate the sensitivity of the outcome to various steps in the process. Sensitivity analysis helps managers, stakeholders and the general public to gain a good overview of the consequences of using different perspectives. This avoids confusion during communication, and it helps to understand if more information or data on certain aspects needs to be collected.

In this case study, the different protection levels for park zoning have been considered as multiple objectives. This allowed to apply a stepwise approach of multicriteria analyses, followed by a multi-objective land allocation, accompanied by sensitivity analysis in various stages of the process. In this way, a complex decision-making process is broken up into digestible pieces that can be analysed, presented and discussed in a transparent manner. Transparency is guaranteed as the content and quality of all data layers can be reviewed and updated at any time. The multicriteria and multi-objective set up can be updated and the analyses repeated with new criteria and indicators, as well as with different weights assigned to them. The relative importance, or weight, of each indicator and criterion is clearly stated, and the sensitivity of the outputs towards the use of these weights can be tested. This helps Park managers to gain a good overview of the consequences of using different perspectives, and decide if more information or data on certain aspects needs to be collected.

To fulfil the need to compare different scenarios for protection level A, the stability of the results when applying different sets of weights for the criteria was tested. It appears that there is no doubt about the units that urgently require a high protection level. Units that contain a relative large contribution of habitat types listed by the EU habitat Directive are quite likely to cover areas of Red List species distributions. Similarly, areas with many outstanding natural features may very well coincide with areas of high importance for biodiversity conservation or relevance according to national or international authorities. An example of this is the attention for high altitude wetlands and nutrient poor wetlands that may harbour various Red List species or have an important contribution to the hydrology (Ramsar 2006). Also the ranks of units with a very low suitability score for protection level A were quite stable when comparing the different sets of weights. Only in the middle range of ranks there was a bit more variation, but the order of magnitude of this variation does not give reason for major discussions on this part of the analyses.

Chapter 5

FOREST LANDSCAPE RESTORATION: APPLYING MCA TO DESIGN AND COMPARE OPTIONS*

5.1 Context

5.1.1 Environmental decision-making problem

Forest resources are experiencing an alarming process of deterioration all over the planet, posing a threat to the protection of global biodiversity. Along with the protection of intact forests, reforestation and restoration interventions have been the main response to the problem of forest degradation and loss. Such interventions, carried out through a number of different techniques, aim at conserving biodiversity and improving both ecosystem functioning and human livelihoods (Lamb et al. 2005). In 2000, IUCN and WWF proposed the Forest Landscape Restoration (FLR) approach as an innovative approach to restore ecological integrity while enhancing human well-being. FLR shifts the attention from the single site to the landscape, as already proposed by Naveh (1994), because redesigning the landscape mosaic can offer opportunities for conserving biodiversity, improving ecological functioning and eventually benefit people. FLR is characterized by few key features: landscape-level view, ecological and socioeconomic perspective, stakeholder involvement and optimization of forest functionality (ITTO 2005).

One of the fundamental issue and a key stage in the FLR approach is the identification of areas that should receive priority for intervention (Vallauri et al. 2005). The prioritization issue is a common topic of conservation science when the shortage of resources calls for the identification of those sites whose protection is likely to provide the maximum benefits (Myers et al. 2000). The selection of an area as a reforestation priority depends on

* The text and graphics of this chapter are based on the following paper: F. Orsi and D. Geneletti, 'Identifying Priority Areas for Forest Landscape Restoration in Chiapas (Mexico): An Operational Approach Combining Ecological and Socioeconomic Criteria', *Landscape and Urban Planning* 94 (1) (2010): 20–30.

the objectives of the reforestation action. Typically, at a landscape scale, multiple and sometimes conflicting objectives have to be considered and traded-off in order to preserve landscape functionalities. Hence, the identification of reforestation priorities is a multi-objective planning problem in which nature conservation and other issues (social, economic) are involved (Kangas and Leskinen 2005). The multi-objective nature of the problem requires the use of different types of variables, often with different levels of (spatial) accuracy. For example, information related to forest cover can be derived from high-resolution satellite imagery, while poverty indices are typically assessed through surveys at municipal or district scale. Thus, suitable approaches are needed that allow handling and integrating such different information (Huby et al. 2007).

5.1.2 Case study description

The study area is located in the central part of the State of Chiapas (Mexico) and includes the *Frailesca* region and surrounding areas. Chiapas shares borders with the Mexican state of Tabasco to the north, Guatemala to the east, the Pacific Ocean to the south and the Mexican states of Oaxaca and Veracruz to the west and north-west, respectively. The study area covers about 18,500 km^2 and falls between 15° 29' and 16° 38' N and between 92° 15' and 93° 48' W (Figure 5.1). Elevation varies greatly, reaching around 2,500 m above the sea level. The area is characterized, in its central part, by a high mountain range (Sierra Madre de Chiapas) with steep slopes of igneous rocks, which follows a north-west–south-east direction. The south-western part is characterized by coastal plains made up of alluvial sediments, whereas the north-eastern one shows plateaus and valleys and the presence of sedimentary rocks. Major rivers follow a north-east–south-west direction.

Forests belong to three main types (Figure 5.1): tropical dry forest (Selva Baja Caducifolia), pine-oak (found in the eastern slopes of Sierra Madre at elevations between 1,300 and 2,500 m) and cloud forest, which is highly diverse, abundant in Chiapas and typical of Sierra Madre at elevations of 1,300 to 2,550 m and high relative humidity (Breedlove 1981; Vega et al. 1999). Two Biosphere Reserves are located along the Sierra Madre and partly included in the study area: La Sepultura in the western side and El Triunfo in the eastern one. La Sepultura covers approximately 160,000 ha and about 24,000 people live there, while El Triunfo has a total surface of about 120,000 ha and a population of about 14,000. Both reserves are part of the Grijalva watershed that supplies water to a number of towns and villages of western Chiapas. The reserves protect some the most ecologically diverse areas of Chiapas and consist of core areas, where human influence is minimized, and buffer zones, where human activities are allowed. The reserves are linked by a

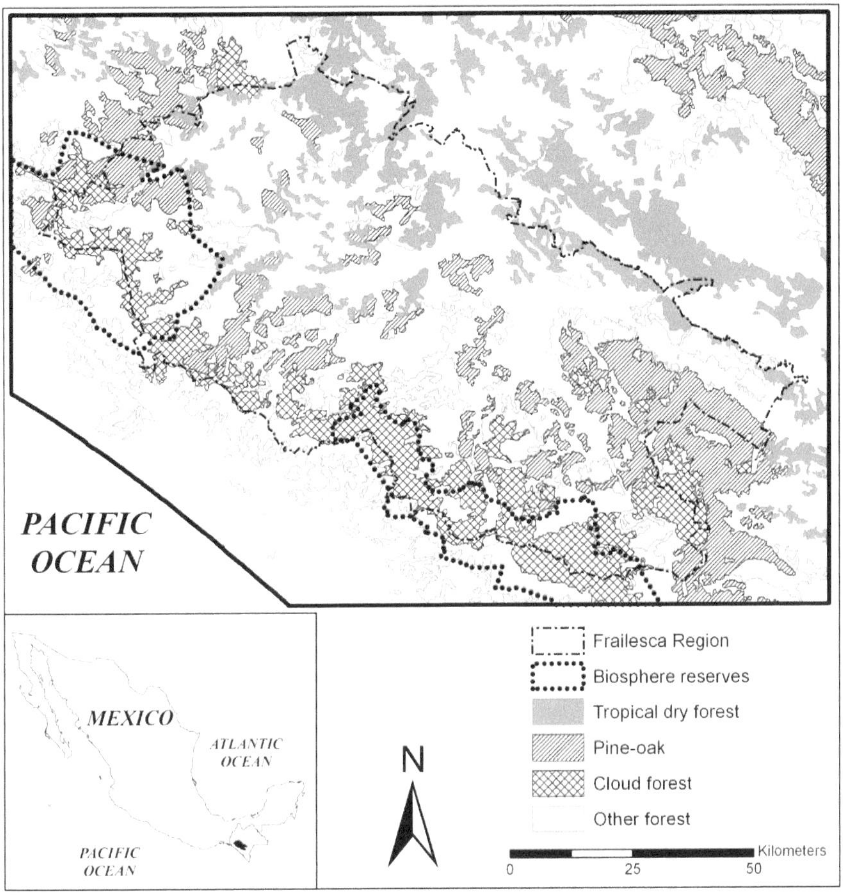

Figure 5.1 The study area in the State of Chiapas, Mexico.

biological corridor, recognized for its importance to species conservation and as part of the Corredor Biologico Mesoamericano.

The biggest urban areas are Jaltenango de la Paz, El Parral, Villa Corzo and Villa Flores, while hundreds of small and very small scattered villages are found across the entire area. According to the analysis carried out by CONAPO (2005), which integrated different socioeconomic information (education, sanitation, etc.) to provide a 'marginalization' index at the municipal level, marginalization in the region ranges from high to very high. Poverty levels are closely linked to the remoteness of human settlements and a strong correlation was found between poverty and the presence of indigenous communities. Three are the main land uses of the region: agriculture, forest and pastures. Major cultivations are corn, beans, coffee, fruits (lemon, mango, papaya) and vegetables (onion, tomato). The actual portion of land dedicated to pastures is varying, as often, especially in the dry season, animals are allowed to trespass on forested or agricultural areas. Land use change is the first threat to forests as agricultural fields and pastures are increased through conversion of forest.

The case study aims at showing an operational application of the Forest Landscape Restoration approach, by identifying reforestation priority areas, and designing landscape-scale reforestation options that aim at improving both ecosystem quality and human living conditions.

5.2 Methods

Previous studies to identify priority areas for forest restoration combined criteria related to biodiversity conservation and disturbance (WCMC 2000; Newton and Kapos 2003), consistently with the idea that priorities can be found in areas where restoration is both needed and likely to succeed. Indeed, several studies maintain that restoration plans should account for the 'restorability' of land (Hobbs and Harris 2001; Suding et al. 2004; Miller and Hobbs 2007). Consistently, the methods adopted in this study are based on the following concept:

$$\text{Reforestation priority} = f(B, F), \tag{1}$$

where:

B represents the need for reforestation: where should biodiversity be protected?

F represents the feasibility of the reforestation process: where is reforestation likely to succeed?

The method was structured into three main steps: generation of suitability maps, design of reforestation options and comparison of reforestation options.

5.2.1 Generation of suitability maps

The following criteria were selected to assess the B factor:

- Distance from ecological corridors: ecological corridors allow animals to move over the landscape and can be considered the core of nature conservation plans (Jongman 1995). Reforestation in or around ecological corridors can improve these structures and thus reduce isolation.
- Distance from existing forest: reforestation of areas in close proximity to existing forests can reduce pressure on such forests, which are reservoirs of native species (WCMC 2000).
- Distance from protected areas: protected areas are a sample of a region's biodiversity to which they provide protection from external threats (Margules

and Pressey 2000). Reforesting in and around a protected site means both enhancing the forested ecosystem and creating a buffer zone that prevents the site from being disturbed.
- Tree species richness: biodiversity hotspots are the main target of a reforestation process aimed at conserving biodiversity.

The following criteria were selected to assess the F factor:

- Distance from agricultural fields: areas around existing agricultural fields are more likely to undergo land use change.
- Distance from roads: roads are a source of disturbance as they allow people to have access to nearby areas.
- Distance from urban areas: towns and villages represent areas of intense human activities, thus demanding large tracts of land in the surroundings in terms of both materials (e.g. timber) and physical space (e.g. land turned into agricultural fields).
- Risk of soil erosion: soil degradation is definitely an issue to be taken into account when thinking about the feasibility of a restoration intervention (Newton and Kapos 2003).

According to these criteria, the most suitable areas to host reforestation interventions are biologically diverse places in or around ecological corridors, forests and nature reserves; not in close proximity to agriculture, roads and settlements; and in areas not exposed to heavy soil erosion. All criterion maps were combined using MCA. A value function was assessed for each criterion (Beinat 1997; Geneletti 2005a) to transform the original score into values between 0 and 1, where 0 corresponds to minimum desirability and 1 to maximum desirability. All value functions, except those for aspect and elevation heterogeneity and for distance from corridors whose shape was concave, were represented by a three-segment polyline (Figure 5.2). The suitability maps for the B and F factors were generated by simply adding up maps of the two groups. Criteria were considered to be equally important within each group (B and F).

5.2.2 Design of reforestation options

The goal was to extract from the suitability maps the best areas up to the fulfilment of the desired reforestation demand, which was set to around 15,000 ha (this is roughly equivalent to the area deforested between 1990 and 2007 in the Frailesca region). The assumption that guided the process of extracting the best areas was that a site can be restored only if it is sufficiently suitable for both B and F factors. Implementing this assumption requires the setting of thresholds that extract most suitable areas from B and F maps. The approach is based

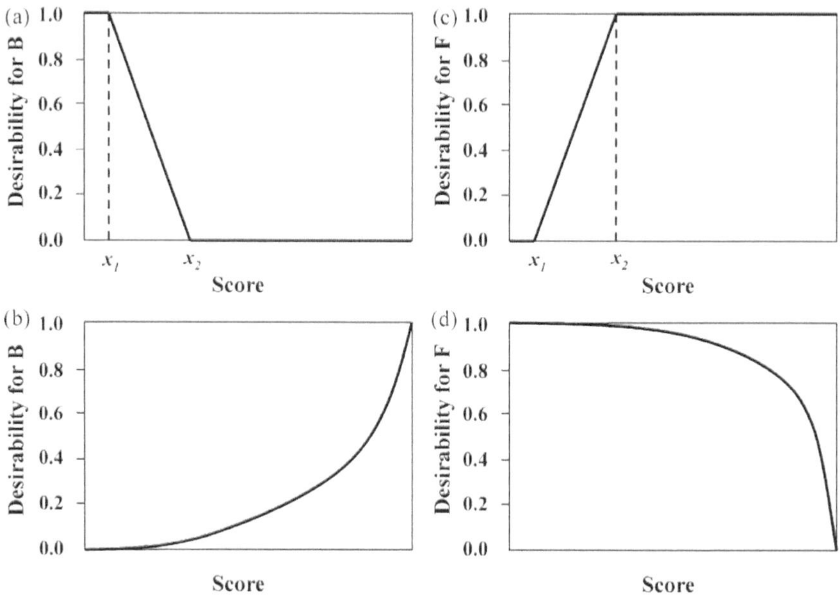

Figure 5.2 Prototypes of the value functions assessed for the eight criteria for factors B and F: (a) for the distance from ecological corridors, the distance from existing forest, the distance from protected areas; (b) for tree species richness; (c) for the distance from agricultural fields, the distance from roads, the distance from urban areas; (d) for the risk of soil erosion.

on the study of histograms of the two maps (B and F) to assess the number of cells above a given threshold. The two maps were crossed on a cell-by-cell basis in order to obtain a combined histogram. Unfortunately, the latter can only provide a frequency information: the number of cells for each pair of values in the suitability maps. The cumulative information was then extracted from the frequency one by considering any pair of values as a threshold pair and computing, for each pair, the number of cells with higher values. A script was written for this purpose with the statistical software R (R Development Core Team 2005). Among all possible threshold combinations, those subtending viable reforestation options were selected on the basis of the following inputs:

- the overall surface to be reforested (reforestation demand);
- the minimum acceptable suitability for B (*bmin*); and
- the minimum acceptable suitability for F (*fmin*).

For each selected combination of thresholds b and f above the minimum suitability (*bmin* and *fmin*) a map was generated, by selecting only cells above both thresholds in the original maps. Contiguous groups of such cells constituted the reforestation sites, whereas each threshold combination shaped a reforestation option, that is, a group of sites. Sites smaller than five hectares were filtered out because they were considered negligible at the landscape scale. Finally, only options with a total reforestation area still above the minimum requirement were considered for further analyses.

5.2.3 Comparison of reforestation options

In order to compare the reforestation options, an additional set of criteria was introduced. These criteria account for the spatial configuration of the options (e.g. patch size and shape), and therefore they can be assessed only after possible reforestation options have been designed. The additional criteria were specifically designed to assess the performance of each reforestation option towards improving the ecological functioning of the landscape and the provision of services to people. They were grouped into ecological and socioeconomic criteria, and are described below, along with the indicator that was adopted to measure them (in brackets).

Ecological criteria:

a. Fragmentation of the landscape (Edge Density)
b. Average compactness of forest patches (Mean Shape Index)
c. Enhancement of ecological corridors (reforestation area occurring within ecological corridors)

Socioeconomic criteria:

d. Land use conversion cost (reforestation area occurring within agricultural fields)
e. Reduction of soil erosion (reforestation area occurring in soil with intermediate erosion risk)
f. Improvement of livelihoods (reforestation area occurring in poorest regions)

The rationale for these criteria sets is that the most suitable reforestation option minimizes landscape fragmentation (criteria a and b), improves ecological networks (criterion c), minimizes conflicts with agricultural land uses (criterion d), contributes to reduce soil erosion in areas where erosion intensity cannot undermine the success of the interventions (criterion e) and contributes to local livelihoods (criterion f). The assumption for the use of the latter criterion was that reforesting poorer regions is likely to improve ecosystem conditions and provision of ecosystem services exactly where people are more dependent on them and more vulnerable to ecosystem degradation. This is the case, for example, of many remote villages along the Sierra Madre mountain range.

Scores were computed that show the performance of different reforestation options against each criterion. These scores were subsequently combined through MCA by using the interval standardization method for criteria a and b, and the maximum standardization methods for the other criteria. Criteria c, e and f were considered as 'benefits' (the higher the score, the higher the desirability), while criteria a, b and d were considered as 'costs' (the higher the score, the lower the desirability). The following are the formulas of the interval functions for benefits (2) and costs (3), and of the maximum functions for benefits (4) and costs (5) (Sharifi et al. 2007):

$$\frac{\text{Score} - \text{Lowest score}}{\text{Highest score} - \text{Lowest score}}, \tag{2}$$

$$1 - \frac{\text{Score} - \text{Lowest score}}{\text{Highest score} - \text{Lowest score}}, \tag{3}$$

$$\frac{\text{Score}}{\text{Highest score}}, \tag{4}$$

$$1 - \frac{\text{Score} - \text{Lowest score}}{\text{Highest score}}. \tag{5}$$

The interval standardization is particularly suitable when a relative scale is used (Sharifi et al. 2007): it is the case of criteria a and b that provide a measure of the ecological improvement brought in by the reforestation. Further, this standardization method emphasizes the differences between alternatives. The maximum standardization instead was selected because of its ability to keep the differences between alternatives: the standardized values are proportional to the original values (Geneletti 2005b). All criteria were considered equally important within the groups, while three weight sets were introduced for the groups, thus accounting for different forest landscape restoration perspectives (balanced, environment-oriented, socioeconomic-oriented), as shown in Table 5.1.

Finally, a sensitivity analysis was performed to test the robustness of the ranking obtained during the choice phase. The analysis considered that both the criterion scores and the weights might be affected by a 10 per cent error, due to the uncertainty in the assessment of value functions and the evaluation of each criterion's relative importance. An iterative simulator

Table 5.1 Weight sets used to compare reforestation options

Group of criteria	Criterion	Perspectives		
		Balanced	Environment oriented	Socioeconomic oriented
Ecological		0.5	0.7	0.3
	a	*0.333*	*0.333*	*0.333*
	b	*0.333*	*0.333*	*0.333*
	c	*0.333*	*0.333*	*0.333*
Socioeconomic		0.5	0.3	0.7
	d	*0.333*	*0.333*	*0.333*
	e	*0.333*	*0.333*	*0.333*
	f	*0.333*	*0.333*	*0.333*

performed some 10,000 iterations to provide information about how rankings are modified because of variations in criterion scores or weights. Further, a comparison of option rankings under different restoration perspectives (balanced, environment-oriented, socioeconomic-oriented) highlighted which options performed well under any perspective and which provided a fluctuant performance. The MCA and the subsequent sensitivity test were implemented using the software package DEFINITE (Janssen et al. 2003).

5.3 Results

The selection of minimum suitability thresholds for both factor B and F led to the identification of 24 reforestation options. However, only 14 options were considered and after that, reforestation sites smaller than 5 ha had been removed (Figure 5.3). Thresholds for B ranged from 0.6 to 0.66, while thresholds for F showed a greater fluctuation, ranging from 0.61 to 0.88. The total reforestation areas ranged from 15059.6 ha (Option 7) to 16141.76 ha (Option 1): a negligible difference considering the overall reforestation area. Figure 5.4 shows the frequency with which cells occur in the reforestation options: black zones represent the reforestation core, that is, the areas shared by all options, constituting a basic option, which ensures high ecological suitability. Most core areas are located in the southern part of the study area, though several smaller patches are found in the north-western sector. The total potential reforestation area, that is, the area of the cells selected by at least one option, is around 28,000 ha of which around one-fourth is represented by cells selected by all options.

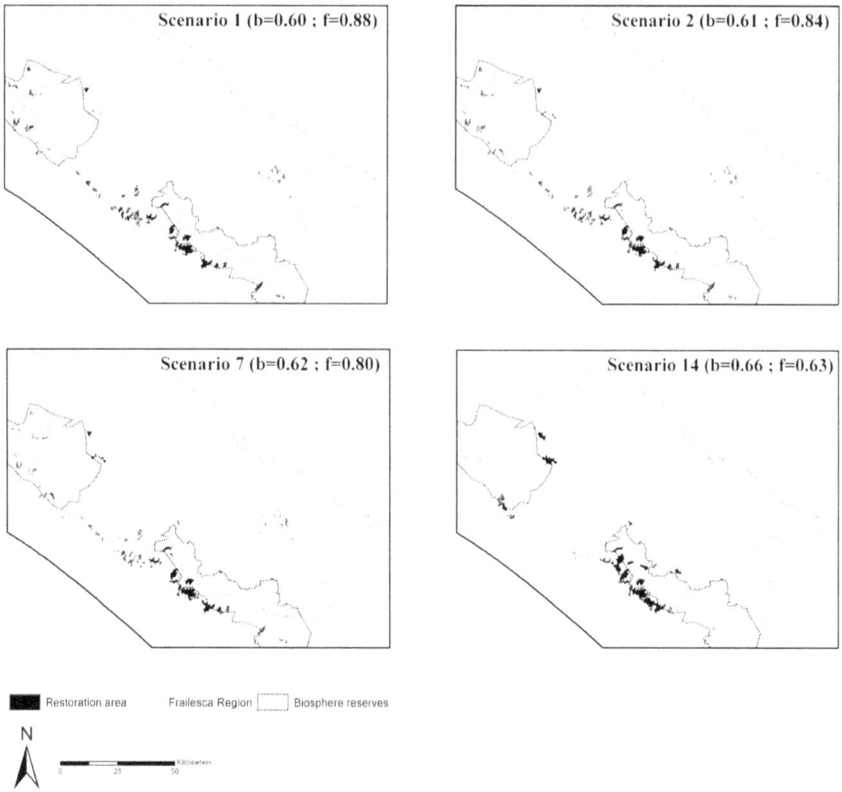

Figure 5.3 A sample of the 14 reforestation options.

Criteria score for all options are presented in Table 5.2. Two observations can be made by looking at the table: the differences among options with respect to the values assumed by criteria a and b are very small, and the value of criterion d is equal to 0 in 7 out of 14 options. The rankings (Figure 5.5) showed that two options perform significantly better than the others under any perspective: Option 9 (0.63 for B and 0.74 for F) and Option 10 (0.64 for B and 0.69 for F). Option 9 is the most preferable option under both the balanced and the socioeconomic-oriented perspectives: in the former case it shows the same performance as Option 10, while in the latter it performs far better than any other reforestation alternative. Under the environmental-oriented perspective it ranks second. Option 10 has the best performance under both the balanced (equal to Option 9) and the environmental-oriented perspectives, while under the socioeconomic-oriented one it ranks third.

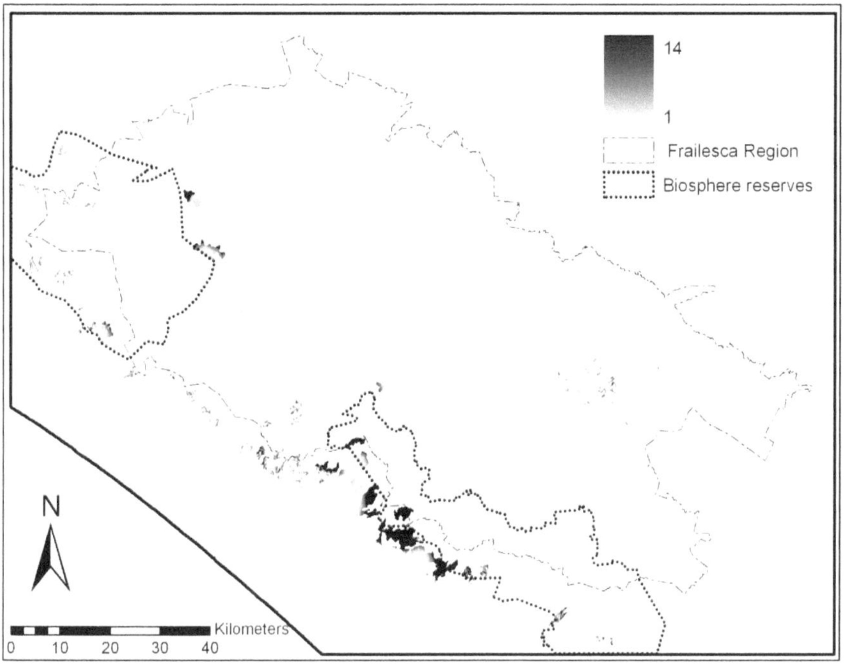

Figure 5.4 Frequency map of the 14 reforestation options (1: area selected by one option; 14: area selected by all options).

What actually makes Options 9 and 10 more preferable is their stability: they rank at the top under all considered evaluation perspectives. When disaggregating the two main components (ecological and socioeconomic criteria), as shown in Figure 5.5, Option 10 is the one that maximizes both components, whereas Option 9 performs very well from a socioeconomic point of view, but only sufficiently from an ecological one. The sensitivity analysis partially accounts for these dynamics when uncertainty in the weights is considered. Option 10, whose total area is 15,338 ha, can be considered the best option (Figure 5.6). The sensitivity analysis carried out for uncertainty in the criterion scores proved that top ranked options are stable under any evaluation perspectives (Figure 5.7). Instead, major fluctuations occur for the mid-ranked options under the balanced and socioeconomic-oriented perspectives and for the bottom-ranked options under the environment-oriented perspective. The uncertainty in the weights affects all options under all perspectives, but the fluctuations are small enough not to modify the rankings.

Table 5.2 Thresholds for B and F and criteria scores of the 14 restoration options

Restoration option	b	f	Area (ha)	Criterion a (dimensionless)	Criterion b (dimensionless)	Criterion c (ha)	Criterion d (ha)	Criterion e (ha)	Criterion f (ha)
1	0.60	0.88	16141.76	5.0007	2.0340	7745.210	0.000	303.700	9720.439
2	0.61	0.84	15849.11	5.0487	2.0001	7541.580	0.000	319.377	9571.473
3	0.61	0.85	15434.62	5.0428	2.0009	7421.285	0.000	303.782	9380.594
4	0.61	0.86	15076.9	5.0401	1.9990	7304.727	0.000	289.892	9206,610
5	0.62	0.78	15861.05	5.0707	1.9834	7409.426	0.000	268.205	9761.377
6	0.62	0.79	15506.75	5.0696	1.9790	7288.157	0.000	262.844	9548.892
7	0.62	0.80	15059.6	5.0697	1.9764	7175.092	0.000	259.920	9233.089
8	0.63	0.73	15940.33	5.0843	1.9806	7509.901	280.640	758.723	9728.480
9	0.63	0.74	15205.48	5.0766	1.9750	7072.261	7.067	629.494	9479.120
10	0.64	0.69	15338.61	5.0123	1.9492	7263.708	288.105	499.290	9265.173
11	0.65	0.64	15675.13	4.9557	1.9938	6983.969	542.908	439.265	9347.779
12	0.66	0.61	15844.07	4.9019	2.0227	7341.928	630.956	398.734	9599.495
13	0.66	0.62	15798.59	4.9022	2.0228	7334.699	623.808	398.409	9560.426
14	0.66	0.63	15671.47	4.8949	2.0281	7330.637	531.536	398.084	9504.218

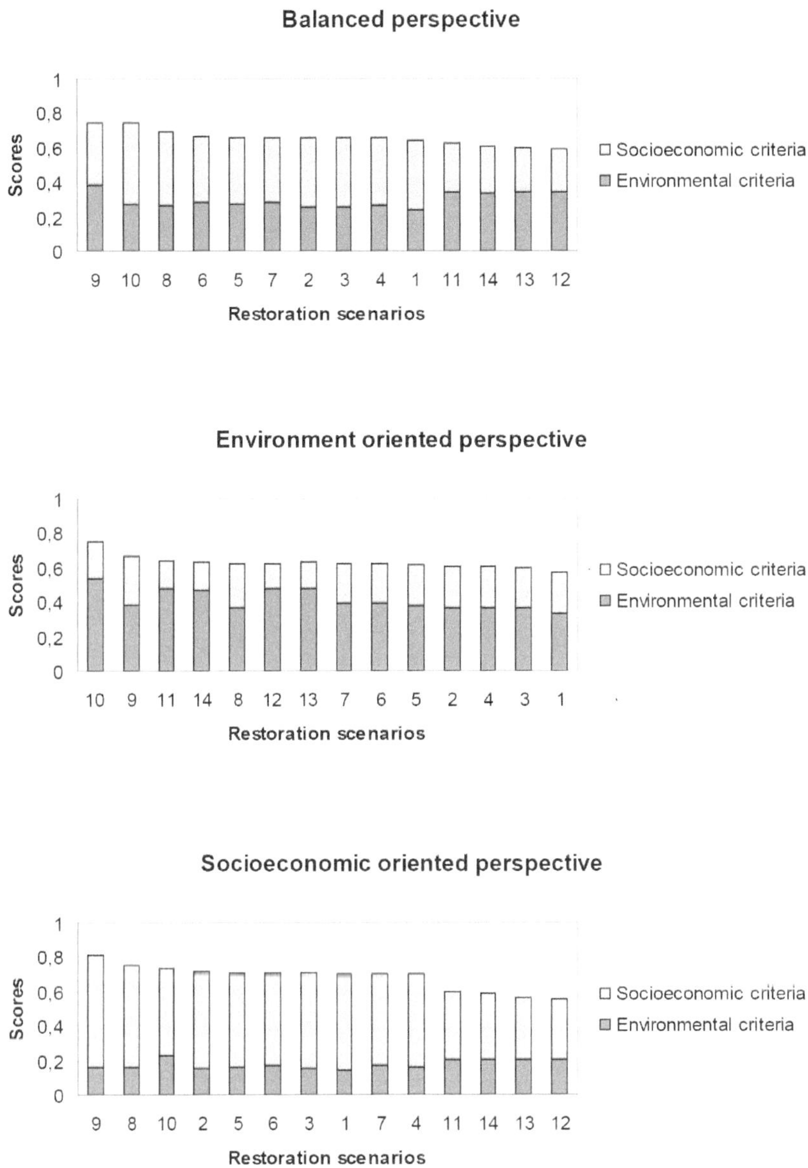

Figure 5.5 Rankings obtained with the three each evaluation perspectives, showing the overall contribution of the environmental and the socioeconomic criteria.

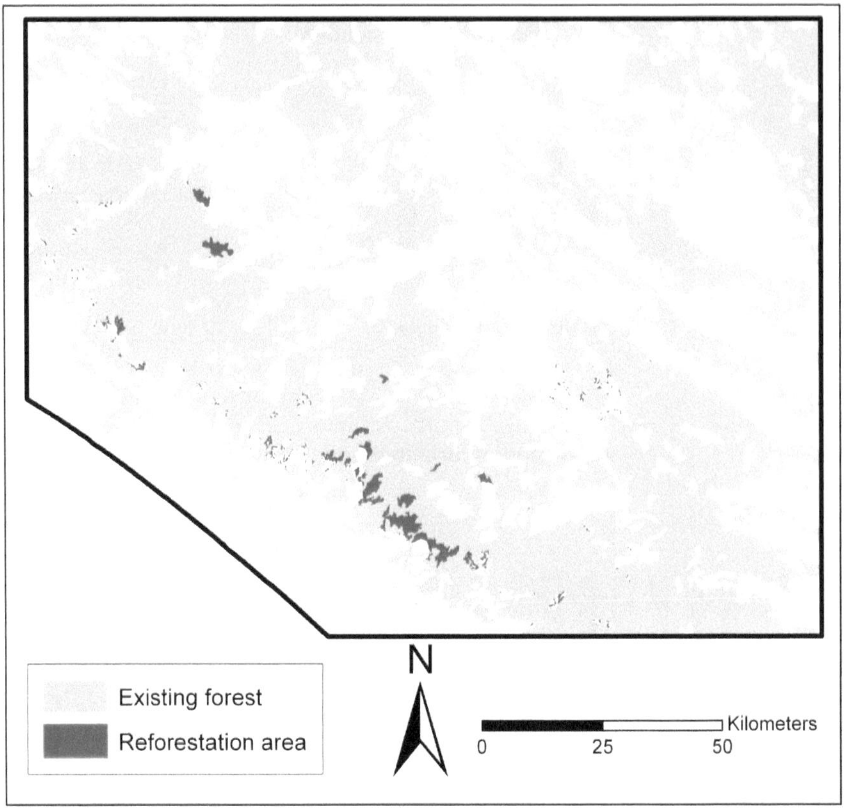

Figure 5.6 The best reforestation option (Option 10), which was obtained with thresholds of 0.64 for B and 0.69 for F.

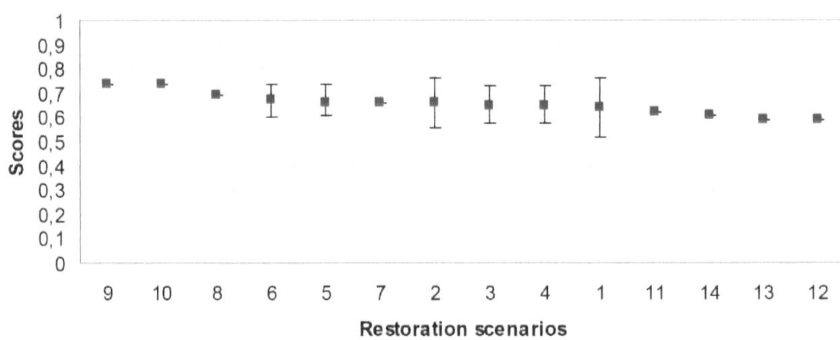

Figure 5.7 Fluctuations in the ranking of the reforestation options under the balanced perspective when uncertainty in the criterion scores is considered.

5.4 Lesson Learned and Conclusions

In FLR, attention should be paid to the landscape mosaic instead of the single forest patch. This concept suggested the methodology adopted in this case study, which moves from the single reforestation site to a configuration of sites through a three-step spatial multicriteria analysis. The need for integrating ecological and socioeconomic objectives was also addressed in subsequent steps: ecological objectives provided a basis for the analysis, while socioeconomic ones allowed for the refinement. This separation was necessary for both making the results ecologically consistent and preventing from a priori excluding valuable reforestation options. The fundamental assumption of the proposed approach is that a site should be accorded priority for reforestation if it offers a potential for biodiversity conservation (B) and if its conditions are likely to make the reforestation intervention succeed (F). This distinction is actually important for dealing with a critical question: should reforestation efforts be directed towards areas most threatened by human activities, or should they kept afar from those areas? To this extent, factor F was interpreted as a cost factor, providing important information about the 'restorability' of a site.

The extraction of the most suitable sites by means of thresholds is a common approach (Riitters et al. 1997; Hirzel et al. 2006), although a risky one, as the land is sliced merely on the basis of cell value: no direct and preliminary control over important factors, say fragmentation, is possible (Geneletti and van Duren 2008). The use of thresholds in this case study was twofold: on the one hand, it separates suitable areas from unsuitable ones; on the other hand, it allowed the non-compensatory (Hwang and Yoon 1981) principle to be implemented. The parameters (land demand, *bmin*, *fmin*) that the methods require to provide a finite set of solutions can be introduced by the decision-maker. Being the minimum thresholds, (*bmin*, *fmin*) the minimum requirements for a cell to become a reforestation priority, their definition is consistent only if the selection of criteria is such that criteria levels are representative of the achievement of objectives those criteria refer to, and if value functions are assessed accordingly. Moreover, land demand and thresholds are obviously related to each other: decision-makers are then able to set a land demand and seek for the highest possible minimum suitability thresholds, or they can fix the minimum thresholds and subsequently establish how much can be reforested depending on the specific context.

The comparison phase shifted the analysis from the site to the landscape: reforestation options were compared as part of the existing landscape. From an ecological point of view, the attention was no longer paid to site suitability, which had already been guaranteed through the first two phases of the methodology, but to the ability of an option to improve landscape connectivity and conservation of key ecological lands (e.g. ecological corridor).

In conclusion, the case study illustrated the application of a method to identify forest reforestation priorities and design landscape-scale reforestation options that aim at improving both ecosystem quality and human living conditions. The method allows decision-makers to generate different reforestation options based on the overall area they are willing to reforest and the minimum ecological suitability level that land should achieve for being considered a priority. Options are then compared to each other through sets of ecological and socioeconomic criteria in order to choose the most efficient one. The use of rather basic GIS operations and the fact that time-consuming computations were not required make the method user-friendly and relatively easy to replicate.

Chapter 6

IMPACT ASSESSMENT OF SKI AREAS: COMBINING GIS INDICATORS WITH MCA*

6.1 Context

6.1.1 Environmental decision-making problem

Environmental impact assessment (EIA) is the process of predicting and assessing the future consequences on the biophysical and human environment of a project, before approval is granted to that project (Canter 1996). The comparison of alternatives (including the do-nothing option, or zero alternative) and the justification of choices are essential elements of any EIA (Geneletti 2014), as requested by EIA legislations around the world. Alternatives can be represented by different project locations, design and technologies, sizes and capacities, ways to construct or operate the project, mitigation measures and so on. EIA being essentially a process to support choices among options by considering multiple types of impacts, the use of MCA can be beneficial. Indeed, an increasing use of MCA has been reported in EIA, also due to the increase in project complexity, data availability and public participation efforts (Janssen 2001).

At a more strategic decision-making level, the process of strategic environmental assessment (SEA) has been introduced to integrate environmental considerations into policies, plans and programmes and evaluate the interlinkages with economic and social considerations (OECD 2006). SEA too requires the analysis and comparison of possible alternative courses of action, and the identification of the most suitable one(s). For these reasons, MCA has become a prominent approach also to support SEA processes, often in combination with GIS (Gonzalez and Enríquez-de-Salamanca 2018). In SEA alternatives are generally of a strategic nature, representing different ways of achieving

* The text and graphics of this chapter are based on the following paper: D. Geneletti, 'Impact Assessment of Proposed Ski Areas: A GIS Approach Integrating Biological, Physical and Landscape Indicators', *Environmental Impact Assessment Review* 28 (2008), 116–30.

the objectives of a policy, plan or programme. For example, the SEA of a spatial plan may compare the effects of different spatial strategies for urban growth (e.g. in Tiitu et al. 2018), or different priorities in land use conversions (see example in Geneletti 2013).

Ski lifts and ski runs are characterized by significant environmental impacts that occur during both the construction and the operation phase, due to activities such as removal of the vegetation cover, surface corrections, piste maintenance and artificial snowing. These activities interfere with fragile high-altitude ecosystems and habitats, and may affect slope stability and water resources. The demand for new skiing facilities has increased steadily in the Alps, fuelled by the reduced availability of snow at existing facilities caused by climate change. This raised concerns about the environmental effects of ski infrastructures.

The presence of ski resorts may significantly affect wildlife by reducing or fragmenting the available habitat, and inducing modifications in animal behaviour. Amo et al. (2007) studied the effect on lizards inhabiting ski slopes and concluded that ski piste construction increased the predation risk of lizards, and consequently affected their body condition. According to Laiolo and Rolando (2005), ski runs produce a negative edge effect: at their edges bird species richness and Shannon diversity are lower than in the forest interior or at the edges of pastures. Watson and Moss (2004) reported the adverse impact on density and breeding success of ptarmigan at a ski area (SA) in Scotland, caused by the influx of predators that followed development. Hadley and Wilson (2004) and Sanecki et al. (2006) maintained that changes in small-mammal proportions are likely to occur as a result of ski centre activity.

The damage to soil and vegetation cover caused by ski piste preparation, and particularly by snowpack compression and machine grading and levelling, are well reported in the literature (Ruth-Balaganskaya and Myllynen-Malinen 2000). Urbanska (1997) concluded that machine-grading of ski slopes causes severe impacts on vegetation, and made spontaneous recovery impossible in the short term. According to Wipf et al. (2005), all types of ski piste management cause alteration of the natural structure and composition of alpine vegetation, leading to a decline in biodiversity. This is worsened by the invasion or introduction of alien species within SAs, which affects native successional series (McDougall et al. 2005; Tsuyuzaki 2002). Gros et al. (2004) reported alteration of soil physic-chemical properties following ski run construction, whereas several other studies addressed the occurrence of soil erosion during piste preparation and use (Mosimann 1985; Ries 1996; Watson 1985).

Geomorphology is affected by ski centres in terms of both destruction of landforms, and interference with processes (Cavallin et al. 1996). The latter may increase natural hazards, such as landslides and avalanches, posing a risk to ski centres, as well as other human assets (Arnaud-Fassetta et al. 2005; Jamieson and Stethem 2002). Degradation of landforms and construction of buildings and facilities affect also the visual quality of the landscape (George 2003). As an example, Kirkpatrick (2004) reported that natural aesthetic conditions of the Kosciuszko National Park (Australia) are extremely poor in the vicinity of ski resorts, and visual quality in the park is also affected by the development of skiing facilities outside the park, but still visible within it. Finally, a particularly critical factor of ski centres relates to artificial snowing. This practice is being increasingly adopted in the Alps, and future climate scenarios suggest that this trend is bound to persist. Artificial snowing increases the input of water and ions within mountain slopes (Stöckli and Rixen 2000), posing a risk in terms of erosion, alteration of hydrologic parameters (hydraulic conductivity, return period of floods, etc.) and changes in local plant community and biodiversity (Rixen et al. 2001; Wipf et al. 2005).

While several papers address the impact of existing ski infrastructures, and monitor their effect over time intervals, only few studies have focused on the preliminary assessment of the suitability of a slope, or a watershed, to host a ski centre (Kliskey 2000; Pfeffer 2003). This type of analysis is useful during the planning process (and associated SEA) to help planners identify broad areas where ski centres should, or should not, be developed, as in the case study described next.

6.1.2 Case study description

Trentino is an alpine region located in northern Italy, and it is a well-known winter tourism destination at both national and international levels. The construction and enlargement of ski centres is a major environmental concern in the region, and in the past more than one project proposal has raised sharp public debate, especially over the interference with glaciers and protected areas. Several EIA have been carried out for ski centres and infrastructures in Trentino since 1989, when EIA was introduced. They represent a good share of all EIA procedures conducted in the region. In order to steer the future development of ski infrastructures, the Spatial Coordination Plan of Trentino identified 30 areas that can be devoted either to the construction of new ski centres, or to the expansion of existing ones. These SAs were proposed without a formal assessment of their environmental effects, because legislation on SEA was still lacking at the time. This study aims at assessing and comparing the impacts of the SAs contained in the Spatial Coordination Plan, so as to provide support to the administration of public authorities during the implementation of the spatial plan, and suggest possible revisions.

Table 6.1 Characteristics of the ski areas (SAs)

Ski areas	Area (ha)	Min. elevation (m)	Max. elevation (m)	Main land cover
SA 1	13	2,200	2,400	High-altitude pasture and grassland
SA 2	130	1,700	2,100	Coniferous forest and high-altitude grassland
SA 3	146	2,000	2,400	High-altitude pasture and grassland
SA 4	113	2,300	2,500	High-altitude pasture and grassland
SA 5	34	1,800	1,950	Coniferous forest
SA 6	33	1,950	2,200	Coniferous forest and high-altitude grassland
SA 7	64	1,700	2,250	Coniferous forest and high-altitude grassland
SA 8	101	850	1,350	Mixed forest and grassland formations

The study focuses on the Fiemme and Fassa Valleys, located in north-eastern Trentino (see Figure 6.1). These valleys are characterized by an economy strongly linked to winter sports. Ski-lifts recorded about 15 million runs per season in the past years, and several international events have been hosted here (e.g. 1991 and 2003 Nordic Ski World Cup). At the time the study was conducted, in the Fiemme and Fassa Valleys there were 11 main ski centres, and the Spatial Coordination Plan provided for the expansion of eight of them, within the designates SAs. SAs cover an overall surface of more than 600 ha, and their elevation ranges between 850 and 2,400 m (see details in Table 6.1).

6.2 Methods

The method is based on the computation of spatial indicators using a GIS to assess critical impacts on the environment, such as ecosystem loss and fragmentation, soil erosion, geomorphologic hazards, interference with flora and fauna, and visibility. Subsequently, MCA is used to generate composite indices, and to rank SAs according to their overall suitability. Finally, the stability of the results is tested through sensitivity analyses.

6.2.1 Impact assessment of ski areas

Figure 6.2 shows the hierarchical tree of the impact factors that was adopted for the assessment. Impacts are organized into six main categories: the first three categories refer to biotic aspects (fauna, forest and nature-conservation

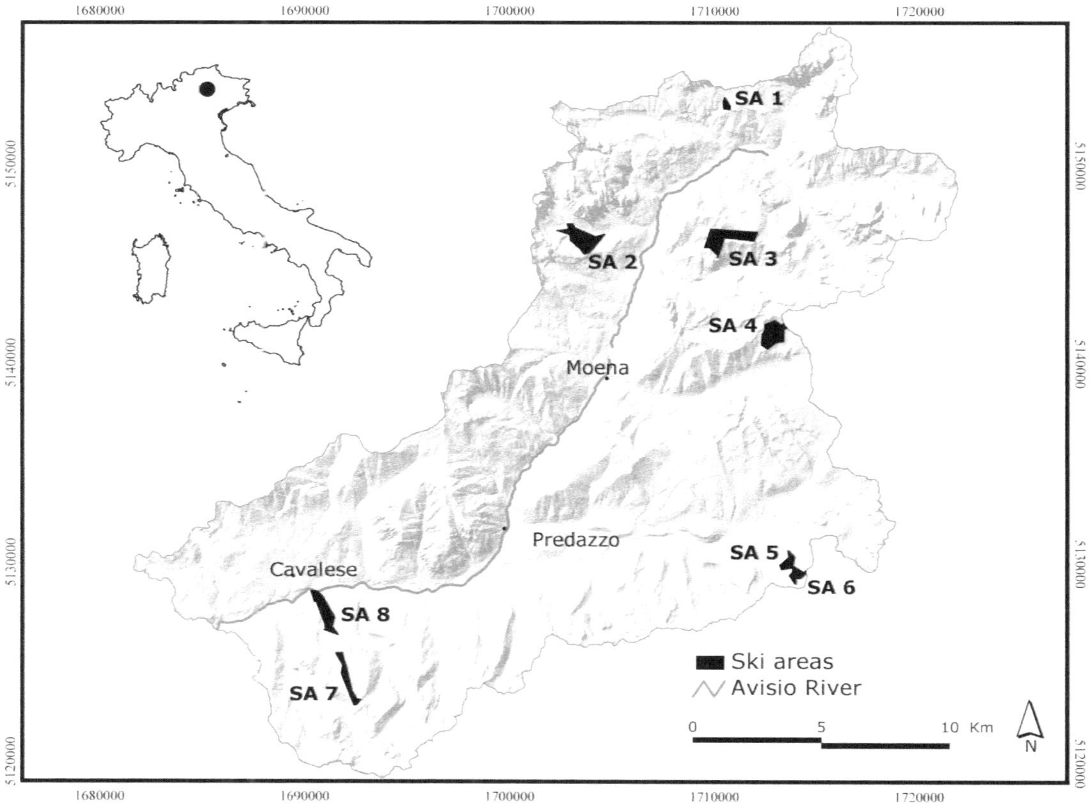

Figure 6.1 Location of the study region in Italy and shaded Digital Elevation Model of the Fiemme and Fassa Valley, with identification of the proposed ski areas.

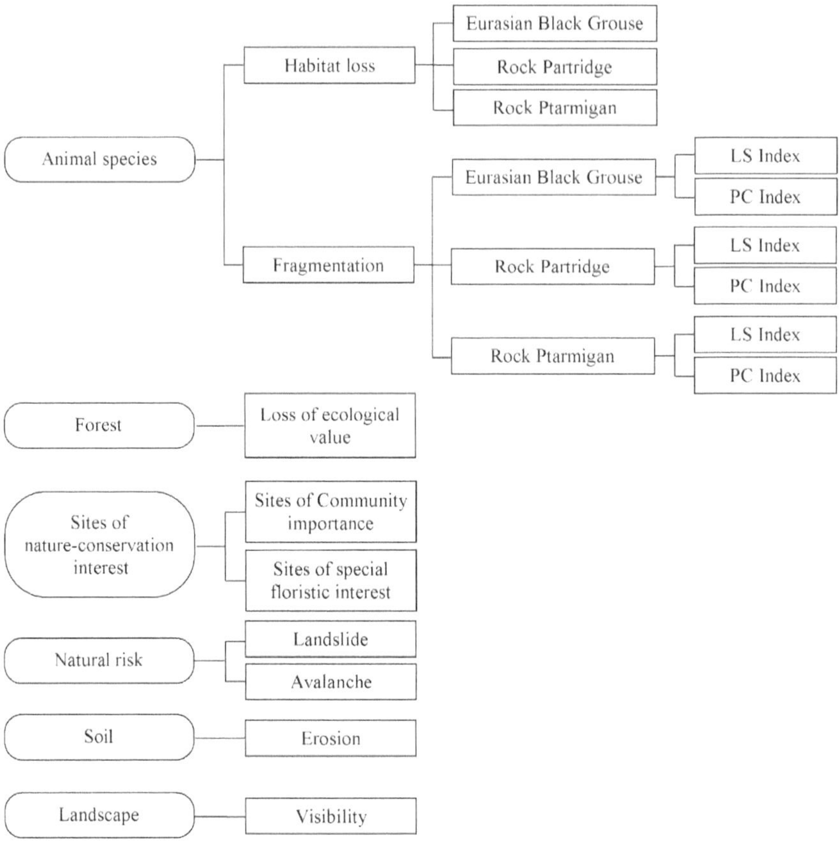

Figure 6.2 Hierarchical structure of the impact factors.

sites), and the remaining ones to geomorphologic hazard, soil and landscape respectively. Impact on fauna was assessed by measuring the expected loss and fragmentation of habitat, which are among the main causes of biodiversity loss worldwide (Noss and Cooperrider 1994). The baseline data were represented by a set of habitat maps provided by the Wildlife Management Plan of Trentino, showing the geographical range of distribution of significant animal species in the study area, namely: Rock Ptarmigan (*Lagopus mutus*), Eurasian Black Grouse (*Lyrurus tetrix*) and Rock Partridge (*Alectoris graeca*). Habitat loss was computed by overlapping SAs (expanded by a 30-m disturbance buffer) with the habitat map of the three species. Subsequently, the area loss caused by each SA was summed up, by applying a weighting factor that refers to the relative importance of the different species. As a result, a habitat-loss impact score was computed for each SA. The weighting factor was assessed by a group of experts who were asked to give each species a biological value.

Habitat fragmentation was assessed by selecting suitable spatial indicators, and then computing their value within the landscape before and after the construction of the project, through simulations. Using measurable indicators offers the advantage of providing a quantitative and objective description of the changes induced by the project. Two indicators that are commonly employed by studies in landscape ecology were selected (Bogaert et al. 2000; Gustafson 1998): the Landscape Shape Index (LSI), and the Patch Cohesion Index (PCI). LSI is a shape index based on perimeter/area ratio that expresses the overall compactness level of habitat patches within a predefined study area (see formula in McGarigal and Marks 1995). PCI measures the physical connectedness of habitat patches within the study area (see formula in McGarigal and Marks 1995). These values increase when habitat patches become closer and more aggregated.

In order to predict the habitat fragmentation caused by the SAs, the two indicators were computed relatively to both the pre- and the post-project conditions, simulating the presence of the SA in the landscape. The landscape of reference to compute the indicators was represented by the watershed that includes each SA. Watershed boundaries were automatically extracted by a Digital Elevation Model (DEM) using a GIS. Impact scores were assessed as the difference between current and future indicator values, for each of the three animal habitats. The software FRAGSTATS 3.3 (McGarigal and Marks 1995) was used to compute the indicators.

The impact on forest was estimated by first assessing the ecological value of the different forest types, and then calculating the surface of each type which is lost to SAs. Literature reviews and consultation with experts belonging to regional research institutes allowed for the generation of a forest evaluation scheme based on the combination of four criteria: naturalness, rarity, outstanding ecological features and disturbances (details can be found in Geneletti 2007a). In the study area, occurrence of plant species is known only at a very coarse scale (5x5 km grid). Detailed surveys and inventories carried out over the past 10 years have generated a comprehensive database, but not a cartographic output with a spatial resolution suitable for this work. However, these surveys were used by scientists belonging to regional research institutes to identify sites of special floristic interest (SSFI). These sites are characterized by the presence of threatened species, which were identified using the regional Red List (Prosser 2001). Additionally, sites of community importance (SCIs) were identified by the local administration in compliance with the European Union 'Habitats' Directive on the conservation of natural habitats and wild fauna and flora (Council Directive 1992/43/EEC). Therefore, maps of SSFI and SCI were used as a reference to assess the impact on nature-conservation areas of proposed ski centres. An impact score was computed by summing up the overall SSFI and SCI area interfered by each ski centre.

Soil erosion induced by SAs was estimated by applying the Revised Universal Soil Loss Equation (RUSLE; Renard et al. 1997):

$$E = R \times K \times LS \times C \times P, \tag{1}$$

Where:

E: Average annual soil loss (ton ha-1);
R: Rainfall-runoff erosivity factor (MJ mm ha-1hr-1);
K: Soil erodibility factor ((ton ha-1)(MJ mm ha-1 hr-1)-1);
LS: Topographic factor (dimensionless);
C: Cover-management factor (dimensionless); and
P: Support practice factor (dimensionless).

Details on how the different factors were computed in the study area are provided in Geneletti (2008b). Geomorphologic risk was assessed by comparing the location of proposed SAs with a map of avalanche and landslide hazard, provided by the Geological Survey of Trentino. Finally, impacts on landscape were addressed by assessing the visibility of SAs within the study region. The visibility of each SA was quantified through viewshed analysis. This analysis, carried out with the GIS software Idrisi Kilimanjaro (Eastman 2003), allows to associate to each cell of the landscape the percentage of a given SA that is visible from there, as a function of terrain topography and land cover.

6.2.2 MCA to compare ski areas

MCA was used to aggregate different impact types into composite scores, and subsequently rank SAs according to their suitability. First of all, the impact scores previously calculated were normalized according to the surface of each SA. In this way, the impact caused per surface unit is taken into account, thus allowing to compare the environmental suitability of the proposed location. At this stage, the actual development (e.g. ski piste, cableway, resort, etc.) has not been designed yet, and therefore the purpose is to assess the plan proposal at a strategic level. The obtained impact scores were used to create an impact matrix and run the MCA.

The standardization of impact scores was performed through the so-called maximum value function (see formula (1) in Chapter 4). This method assumes that a linear relationship exists between the impact scores and the perceived

Table 6.2 Weight set of the four perspectives

	Perspective			
	Nature conservation	**Geomorphology**	**Landscape**	**Balanced**
Animal species	0.271	0.091	0.100	0.167
Forest	0.271	0.091	0.100	0.167
Site of nature-conservation interest	0.271	0.091	0.100	0.167
Soil	0.076	0.346	0.100	0.167
Geomorphologic hazard	0.076	0.346	0.100	0.167
Landscape	0.035	0.035	0.500	0.167

significance of the impacts. This means that the same increase of an impact score is equally valued all over the range of the possible scores. This was considered suitable to the case study because no additional information was available about critical impact thresholds. Furthermore, this method offers the advantage of keeping the ratio between the original impact scores and the standardized ones.

Weighting represents a critical stage aimed at including into the analysis the viewpoints of decision-makers and stakeholders. In this study, four different perspectives were simulated by using different weight sets. Three of these perspectives are characterized by a very strong preference towards one environmental aspect (nature, geomorphology and landscape, respectively), whereas the fourth one simulates a balanced point of view (Table 6.2). Different perspectives were introduced only for the weights of impacts of the first hierarchical level, whereas at sublevels equal weights were assigned. Subsequently, the impact scores of each SA were aggregated through weighted linear combination. This method calculated a composite impact score for each area by first multiplying the standardized impact scores by their weight, and then summing the weighted scores for all impact types. Through weighted linear combination, each SA was given a composite score ranging between 0 (worst performance) and 1 (best performance).

Sensitivity analysis was applied to assess the stability of the results with respect to changes in the input factors. First, an analysis was conducted to account for uncertainty in the impact scores. To this purpose, perturbations were imposed to the original scores, and their effect on the final results were studied. This was achieved by setting a variation range for each score, generating a random number within that range and performing the weighted linear aggregation with the new scores. The procedure was repeated 10,000 times to get a statistically relevant result. A variation range equal to ± 20 per cent of the original score was assigned to those impact types that were assessed by comparing the location of SAs with existing thematic layers (e.g. habitat maps, forest map, hazards map). A variation range equal to ± 30 per cent was considered for the impacts whose assessment required the use of models carried out in this study (e.g. fragmentation, soil erosion). The rationale behind this choice is that the first category of impacts relies on information and data already validated and agreed upon, whereas the second one relies on experimental analyses, which therefore are more prone to errors and inaccuracies. A second sensitivity analysis was run to test the effects caused by changes in the weights assigned to the different impact types. The performance of the SAs was computed for all possible values of a selected weight. During this analysis, the ratios between the remaining weights are kept constant (Janssen et al. 2001).

6.3 Results

Some examples of the results produced during the impact assessment stage are described next (see Geneletti 2008b for further details). Figure 6.3 shows the fragmentation impact scores of the different SAs referred to the Rock Ptarmigan habitat. The impact of SA 8 and SA 5 are not represented in this figure, because they are located within the same watershed of SA 7 and SA 6 respectively. Figure 6.4 shows the forest value map, which was overlaid to the interference buffer of the different SAs: impact scores were computed by summing up the value of the affected forest areas. Figure 6.5 illustrates the visibility map for SA 3. Viewshed analysis was conducted within a 5-km radius. That means that each cell within this range is given a visibility value: the higher the value, the more visible the SA is from that location. Subsequently, visibility maps were aggregated into impact scores by summing up the cell values.

Through the impact assessment stage, an impact score was associated to each SA, and for each of the impacts presented in Figure 6.2. Those scores were then normalized to produce the matrix shown in Table 6.3. Weighted summation of the normalized scores was repeated for the four perspectives, obtaining the results shown in Figure 6.6. As it can be seen, SA 1 and SA 8 are the least-impacting ones under three out of the four perspectives. However, the performance of SA 1 drops when high relevance is given to landscape aspects. In this case, the best performing area becomes SA 4. At the bottom of the ranking, SA 6 is considered to have the highest environmental impact according to two perspectives, but its performance improves when importance is given to geomorphologic hazard.

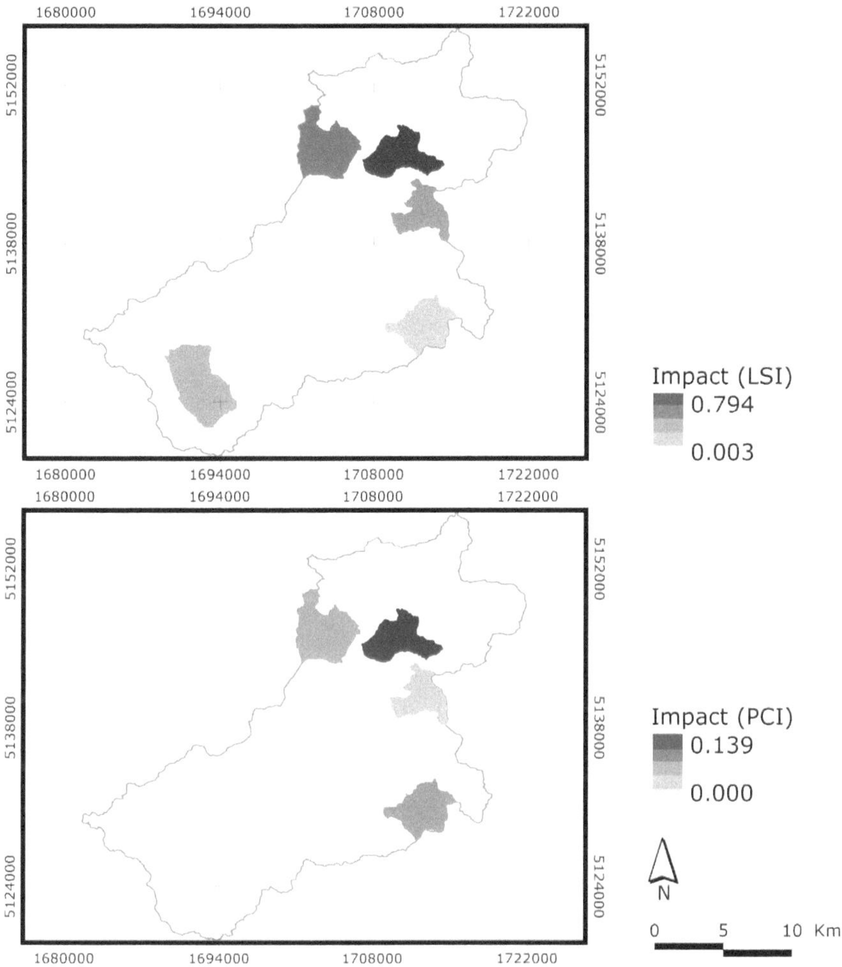

Figure 6.3 Fragmentation impact on Rock Ptarmigan habitat within the ski area watersheds (PCI: Patch cohesion index; LSI: Landscape shape index). Values express the difference between current condition and future scenario.

The results of the sensitivity analysis conducted for all the four perspectives were summarized in frequency tables, which show how many times an area ranks in every position. As an example, the results for the nature-conservation perspective are presented in Table 6.4. This table shows that SA 8 is considered the least impacting area (i.e. it holds the first position in the ranking) in over 90 per cent of the iterations. That means that its performance is stable with respect to variations in the impact scores. On the contrary, the performance of the SAs in the mid-sector of the ranking

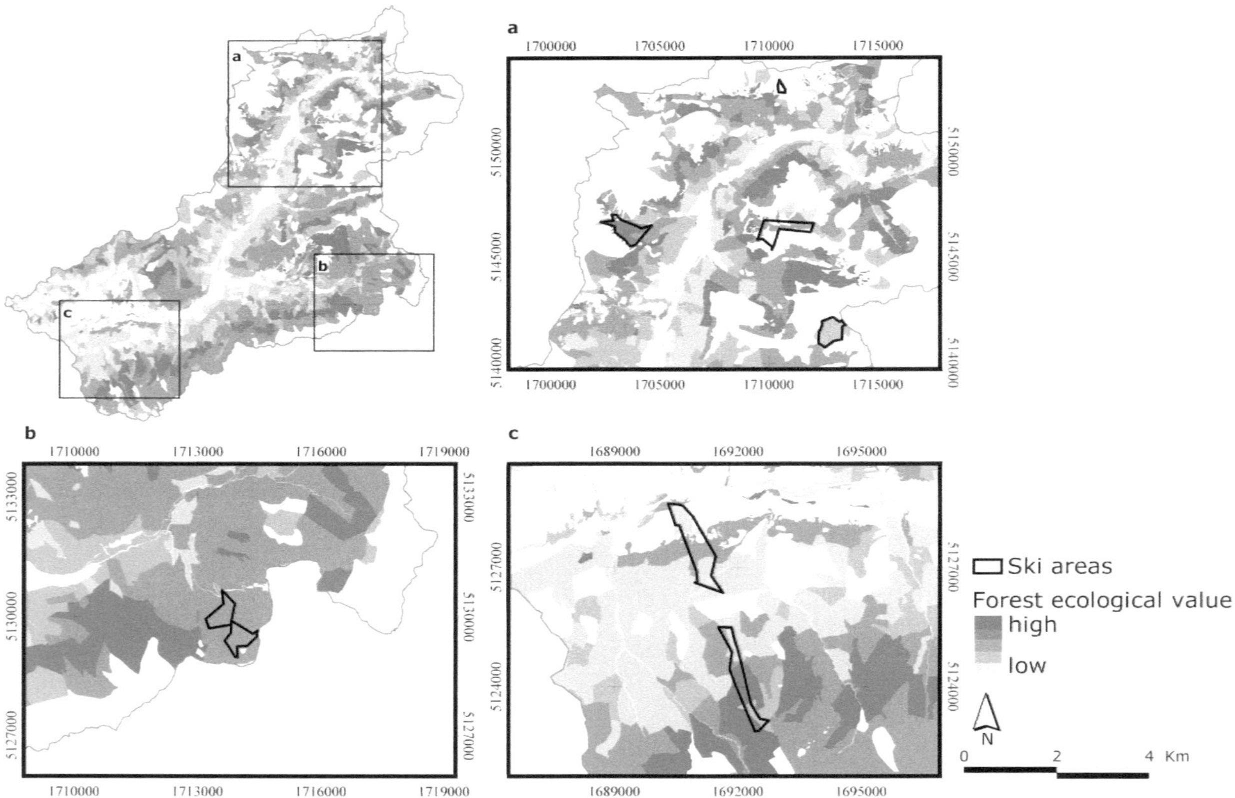

Figure 6.4 Ecological value map of forest areas (simplified legend for b/w representation).

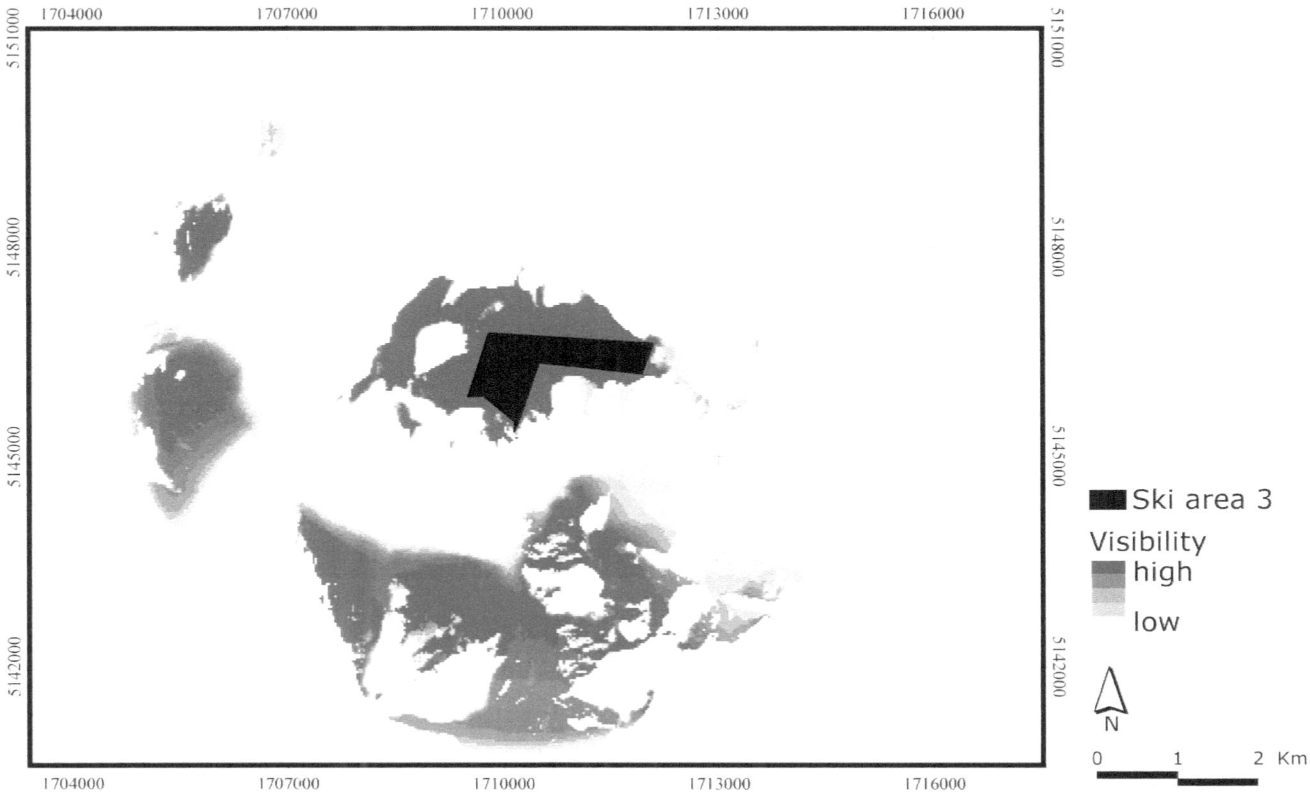

Figure 6.5 Viewshed analysis for ski area 3 (simplified legend for b/w representation).

Table 6.3 Impact matrix with normalized scores

	SA 1	SA 2	SA 3	SA 4	SA 5	SA 6	SA 7	SA 8
Animal species								
Habitat loss								
Eurasian Black Grouse	0.99	0.25	0.00	0.02	0.67	0.17	0.60	1.00
Rock Partridge	0.22	0.63	0.00	0.01	0.98	0.45	1.00	1.00
Rock Ptarmigan	0.86	0.86	0.10	0.00	1.00	1.00	0.79	1.00
Fragmentation								
Eurasian Black Grouse								
LS Index	1.00	0.20	0.00	0.22	1.00	0.75	0.75	1.00
PC Index	1.00	0.87	0.00	0.99	0.99	0.70	1.00	1.00
Rock Partridge								
LS Index	0.81	0.94	0.00	0.27	0.79	0.84	1.00	1.00
PC Index	0.99	0.97	0.00	0.99	0.71	0.57	1.00	1.00
Rock Ptarmigan								
LS Index	0.81	0.99	0.85	0.00	1.00	1.00	0.90	1.00
PC Index	0.81	1.00	0.00	1.00	1.00	1.00	0.42	1.00
Forest								
Ecological value	1.00	0.00	0.57	0.30	0.21	0.17	0.16	0.50
Sites of nature-conservation interest								
Sites of community importance	1.00	1.00	0.97	1.00	0.00	0.00	1.00	1.00
Sites of special floristic interest	0.00	1.00	0.94	1.00	1.00	1.00	1.00	1.00
Soil								
Erosion	0.39	0.00	0.09	0.66	0.55	0.92	0.12	0.42
Natural hazard								
Landslide	0.72	1.00	0.94	0.00	1.00	0.77	0.76	0.94
Avalanche	1.00	0.91	0.55	1.00	1.00	0.00	0.77	1.00
Landscape								
Visibility	0.00	0.89	0.86	0.91	0.46	0.43	0.66	0.65

tends to fluctuate more. Note that in this table the values of each row do not necessarily sum up to 100 per cent because when two areas rank in the same position the next position is not assigned. The actual fluctuation in the SAs composite impact scores obtained with the above-described analysis is presented in Figure 6.7. It is clearly visible that, even though the mean impact score of SA 1 and SA 8 is very similar, the two SAs behave differently when uncertainties are introduced in the impact scores: SA 8 is stable, whereas SA 1 proved to be sensitive to uncertainty factor.

Table 6.4 Frequency table of the nature-conservation perspective obtained by introducing uncertainty factors in the impact scores

	1st	2nd	3rd	4th	5th	6th	7th	8th
SA 8	91	8	0	0	0	0	0	0
SA 1	6	48	23	11	7	4	2	0
SA 4	3	24	24	13	11	11	11	3
SA 7	0	4	18	36	31	11	1	0
SA 2	0	11	23	22	17	15	11	1
SA 3	0	5	11	13	15	2	29	7
SA 5	0	0	1	6	19	39	35	1
SA 6	0	0	0	0	0	1	11	88

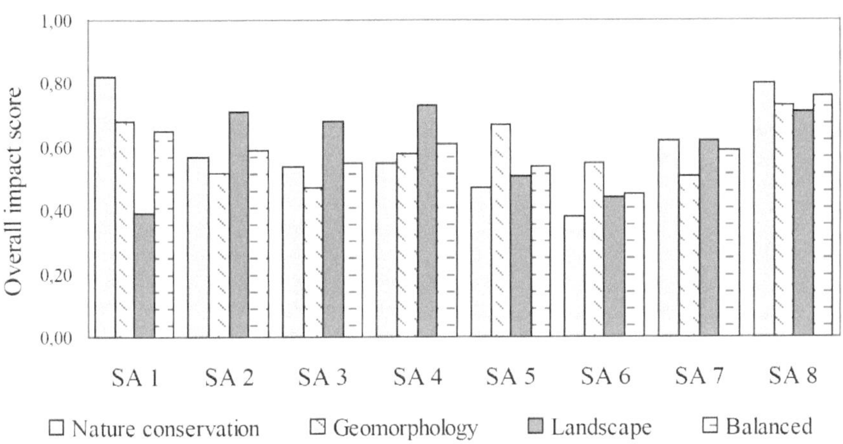

Figure 6.6 Composite impact score of ski areas under the four perspectives.

By way of example, the results of the second sensitivity analysis (conducted to test the effects caused by changes in the weights assigned to the different impact types) are illustrated Figure 6.8. This figure shows the results of the analysis conducted for the weight of the impact on forest. On the x-axis is represented the range of the possible weight values (i.e. from 0 to 1), and on the y-axis the composite performance score of the SAs, which are represented by the sub-horizontal lines. When two lines cross, a rank reversal between the relevant SAs occur. The small circle in Figure 6.8 highlights a rank reversal between SA 1 and SA 8. The vertical dashed line indicates the original weight value. Also in this case, the figure represents the results obtained with the nature-oriented perspective. SAs not represented in this figure lie entirely below SA 4 in terms of performance. This means that, irrespective of the

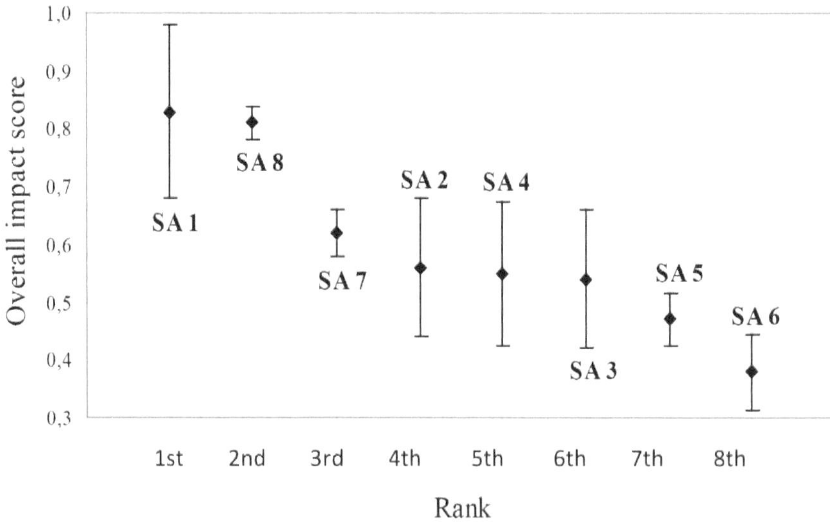

Figure 6.7 Fluctuation of the composite impact scores obtained introducing uncertainty factors in the individual impacts.

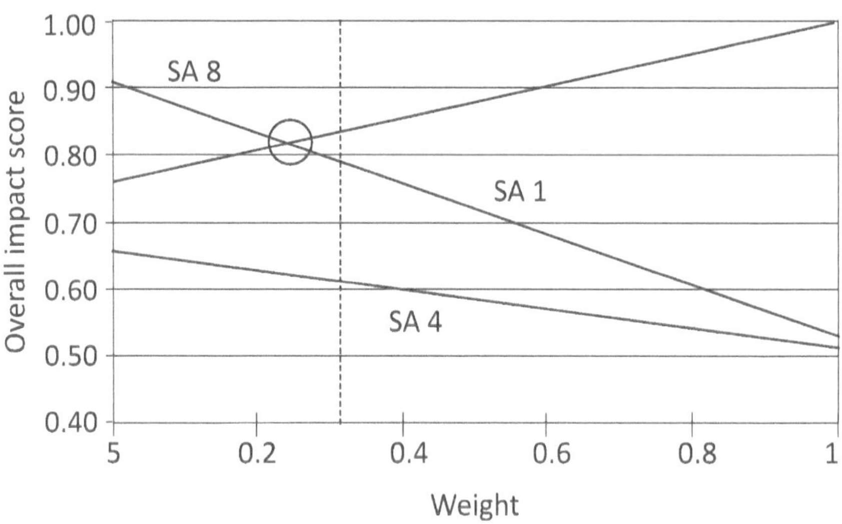

Figure 6.8 Illustration of how the composite impact score of Ski Areas 1, 4 and 8 changes according to the weight assigned to the impact on forest. The weight is represented on the x-axis, and the overall score of the three alternatives on the y-axis. The three lines represent the trend of the scores of the alternatives for all possible values assigned to the weight. This representation allows us to visualize the reversal points (see small circle) corresponding to the weights that cause a reversal in the rank order of the alternatives.

weight assigned to the impact of forest, their performance is lower than the one of the SAs represented in the figure. Among the weights of all impact factors, the one for forest (represented in Figure 6.8) turned out to be the most sensitive, that is, the one where rank reversals occur closest to the original weight. According to this analysis, there are two areas that outperform the others: SA 1 and SA 8. On the contrary, SA 5 and SA 6 perform constantly at the bottom of the ranking.

6.4 Lesson Learned and Conclusions

This study applied a methodology to compare the suitability of the SAs proposed by a spatial plan. The purpose was to assess the impact of the plan's provisions in a context in which SEA is not active yet, and therefore the plan did not undergo a formal evaluation of its environmental effects. Therefore, the method consisted in the assessment of the overall suitability of the areas, rather than of specific ski infrastructure projects. For this reason, SAs were compared in terms of impact per surface unit, so that larger areas are not systematically penalized. The proposed method aims at providing a support to public administration during the implementation of the spatial plan, by suggesting possible revisions, as well as rankings of development priorities. These rankings refer only to the environmental impacts of SAs, and therefore need to be complemented by an evaluation of the expected socioeconomic effects. In this way, a complete overview of the benefits of the development of the different SAs can be gained and used to devise a strategy for the actual implementation of the plan.

Concerning the case study results, the study concluded that, from an environmental standpoint, at least two of the proposed SAs (SA 5 and SA 6) are affected by major drawbacks. The poor performance of these areas was confirmed by sensitivity analyses, which tracked the effect of uncertainty factors affecting data and methods (uncertainty on the impact scores), as well as evaluation perspectives and decision-makers' priorities (uncertainty on weights). As shown in Figure 6.1, SA 5 and SA 6 lie next to each other, and represent two proposed enlargements of the same ski centre. They are located at high altitudes and within an SCI, hence the significant impact. Their main land cover is represented by *Pinus mugo* bushes (listed as a priority habitat in the EC Habitats Directive), larch and spruce forests, and montane grass formations. The findings of this study suggest a revision of the provisions of the Spatial Coordination Plan concerning further ski centre development within these areas.

SA 1 and SA 8 are the best performing areas in environmental terms, and this can be used as an indication to steer the implementation of the Plan.

SA 8 proved to be stable with respect to uncertainty factors, and ranked in the first two positions independently from the perspective adopted. SA 8 is located at a lower elevation with respect to the other SAs, and it interferes with less sensitive ecosystems, and slopes less prone to soil erosion and instability. SA 1 alternates in the first two positions with SA 8, but its performance decreases dramatically in the landscape perspective. This is because SA 1 is located in an open, hence visible, geomorphologic context. In general, SA 1 is characterized by high sensitivity to uncertainty factors related to both weights and impact scores. This suggests that the implementation of SA 1 should be carried out carefully, by paying special attention to the location, size and characteristics of the ski infrastructure projects.

The approach presented in this case study aimed at being extendable to the whole regional territory. For this reason, it relied on data that are routinely acquired and updated by the regional administration, or that can be generated from such data with relatively little efforts. GIS layers were based on primary data extracted from official documents (e.g. Wildlife Management Plan, Forest Management Plan, etc.) to ensure that they were quality tested and validated. Evaluation criteria and value assessments (e.g. relevance of animal species) were derived by previous studies that involved both scientific experts and public administration officers. In this way, the impact assessment is consistent with other nature-conservation approaches and policies adopted at the regional level.

Finally, the method can be extended to other contexts, particularly where SEA regulations are still lacking, and therefore there is a need for a set of indicators suitable for impact assessment above the project level. At this level, the use of GIS modelling is efficient because it allows the carrying out of rapid analyses at landscape scale. Specific field surveys can be integrated at a later stage, for example, to design the most suitable alignment for a ski piste or suggest mitigation measures.

CONCLUSIONS

The chapters of this book illustrated how MCA can promote a logical and well-structured decision-making process, overcoming informal judgements and decisions unsupported by analysis. For example, the case study on protected area zoning (Chapter 4) described how a complex problem can be broken down into smaller pieces, which are first addressed through separate MCA, and then combined. Similarly, the case study on forest landscape restoration (Chapter 5) adopted a stepwise approach to first apply MCA to narrow down the problem to a set of possible solutions, and then compare them through additional, and more specific, criteria. MCA helps also to combine factual information (e.g. results of the impact modelling) with value-based information (e.g. stakeholders' views) in an explicit way. The case study in landfill site selection (Chapter 3) illustrated the role of these two types of information throughout the decision-making process: from the first screening of suitable areas, to the comparison and ranking of possible sites.

MCA can contribute to make environmental policy and management decisions more transparent and justified, by improving problem structuring, exploring the view of the stakeholders and enhancing their understanding of the problems and the issues at stake (Kiker et al. 2005). As Gamper and Turcanu (2007) put it: 'All in all, the added value of MCA for public decision making clearly lies in its ability to reveal preferences, in a more direct and practical way than other decision-support tools.' Nevertheless, MCA is sometimes perceived as a technocratic approach, and identified with the computation undertaken to combine criteria scores and produce a ranking of alternatives. However, this is a narrow interpretation of MCA, as the methodological insights and case studies presented in this book showed. Uncovering, and exploiting, the full potential of MCA to support decision-making requires a broad understanding of decision processes, the knowledge of other methods and tools that can be combined with MCA and appropriate skills and resources, as described next.

Any assessment that involves multiple criteria and objectives inherently requires some form of weight assignment. This is often the most controversial and disputed part of the analysis, but it is also the moment that offers the greatest potential for stakeholder engagement and public participation in a way that is meaningful and influential. One of the great advantages of using MCA is that it makes the weighting process transparent: weights are assigned in a specific moment of the process, they are clearly stated and the way in which they are used to compare the alternatives is made explicit. Additionally, their role in determining the final results can be easily unveiled through sensitivity analysis. For example, the analyses conducted for the case study on the impact assessment of sky areas (Chapter 6) allowed understanding and visualizing how the results change, according to different perspectives on the relevance of the different impact types.

Sensitivity analysis offers the opportunity to stimulate discussion, better understand and address conflicts related to different viewpoints and value systems among the actors (Gamper and Turcanu 2007). In turn, this often leads to higher agreement and more effective search for mutually acceptable solutions. Obviously, the ultimate choice about who is entitled to elicit weights (e.g. decision-makers only? Stakeholders' representatives? The general public?), and whose opinions count most is ultimately a political issue, and depend upon the specific decision context (DCLG 2009). However, it is important to note that we do not have necessarily to reach a consensus on weights before proceeding with an MCA. In many situations, it may be useful to run separate MCA analyses in parallel, by using the opinions of different groups of actors, and compare the results. Only then may one decide whether more consultation processes are needed, or whether the information is enough to reach acceptable results.

MCA offers a flexible framework for decision support and can be combined with other approaches and tools that are often relevant for environmental decision-making, such as GIS and participatory approaches. All the environmental decision-making problems presented in the case studies of this book have a spatial nature. Hence, the use of GIS necessary to easily manage the required input data, and produce suitable outputs. MCA in a GIS environment (or spatial multicriteria analysis, SMCA) takes advantage of the capability of GIS to manage, process and represent spatial information, as illustrated, for example, by the land suitability maps for landfill siting and Park zoning (Figures 3.2 and 4.3, respectively), and by the impact maps of ski areas (Figures 6.3–6.5). Stakeholders play a vital role in MCA, as they are expected to provide inputs throughout the process. For this reason, MCA is often used together with participatory approaches and methods to elicit stakeholders' opinion. For example, in Chapter 3 actors were engaged first by conducting a broader stakeholder analysis, aimed at identifying and assessing criteria to be considered for landfill siting, and then by interviewing experts, in order to get more specific and technical insights on criteria that should be used to compare potential sites. Other examples of methods that can be coupled with MCA include Delphi surveys, focus groups, workshops and questionnaires, as described in the review of MCA in nature conservation decision-making presented in Chapter 2 (see Figure 2.3).

In terms of tools for running MCA, several software packages are available (see, e.g., the comparison of MCA software by Mustajoki and Marttunen (2017)). They differ according to the specific MCA method that they implement and the specific stage that they focus on. Often, the analyses required to run MCA are relatively simple and can be implemented in a simple spreadsheet (or GIS, in case spatial criteria are used), so software availability is hardly a constraint. However, users may benefit from tailored-made software, particularly in terms of graphical and visualization features, friendly interface and routines for the most sophisticated computations, such as sensitivity analysis. These features are particularly important if interaction with stakeholders is foreseen, such as during workshops, open meetings or focus groups.

Even though in most MCA methods the operational stages are relatively simple and can be learned quickly, successful application of MCA requires skills and experience in the broader field of decision analysis. Users need to understand and manage critical issues, such as, for example:

- the role that the different actors play during the decision-making process (Are they involved in the right moment? Are they providing the required input?);
- the selection of the criteria to be included in the analysis (Are they linked to decision objectives? How? Are they representative of the issues of concerns? Do all actors understand them in the same way?);
- the implication of different scoring and weight assessment techniques;
- the choice of the most adequate method for aggregating criteria (and the implications that different methods may have for the results); and
- the selection of meaningful sensitivity analyses to explore the results comprehensively and test their stability and so on.

Additionally, the application of MCA requires skills as a facilitator and communicator, as the analysts are involved also in the collection of stakeholders' opinions (surveys, interviews, etc.), and in the presentation of the results (e.g. during public debate).

Concerning the resources needed to implement MCA, they obviously depend upon the specific problem. Costs are not related to the application of the tool as such, but to the fact that an MCA approach requires deep interaction between the MCA specialist and the other actors involved in the decision. The mere computations needed to run MCA and achieve a ranking of alternatives may take just few minutes. In most situations one MCA specialist is sufficient to perform the analysis, possibly jointly with another person with expertise in stakeholder interaction and communication. This obviously refers to the MCA analysis per se, and does not include the data collection and processing required to generate the input data. Stakeholder and public engagement activities usually turn out to be the most expensive and time-consuming activities. However, it can be argued that these activities are not related strictly to MCA, but to the decision-making process in general. Good MCA calls for the analysts to be involved since the beginning of the process, to help also during problem structuring and design, often the most critical stages.

In conclusion, MCA allows criteria to have considerable mutual difference in nature, and provides a framework to incorporate stakeholders' values in a transparent way. These two features make MCA particularly appealing to support complex decision problems, characterized by many and often conflicting objectives, the need to broadly engage stakeholders with different priorities and the fact that it is not possible to find an alternative that is the best in terms of all perspectives. Hence, good performance on some criteria needs to be traded-off against poorer performance on other criteria. This is a typical condition of environmental decision-making problems, where objectives are multiple by definition, and where trade-offs are an unavoidable and inherent aspects of the process, as the case studies presented in this book demonstrated. It is my hope that the book can provide guidance to practitioners, researchers and decision-makers on how to design and apply MCA approaches to effectively address environmental challenges.

ANNEX

References of the MCA studies reviewed in Chapter 2:

1. Ferretti, V. and Pomarico, S. 2013b. 'Ecological land suitability analysis through spatial indicators: An application of the Analytic Network Process technique and Ordered Weighted Average approach'. *Ecological Indicators*, 34: 507–19.
2. Ferretti, V. and Pomarico, S. 2013a. 'An integrated approach for studying the land suitability for ecological corridors through spatial multicriteria evaluations'. *Environment, Development and Sustainability*, 15: 859–85.
3. Bottero, M., Comino, E., Duriavig, M., Ferretti, V. and Pomarico, S. 2013. 'The application of a Multicriteria Spatial Decision Support System (MCSDSS) for the assessment of biodiversity conservation in the Province of Varese (Italy)'. *Land Use Policy*, 30: 730–38.
4. Newton, A. C., Hodder, K., Cantarello, E., Perrella, L., Birch, J. C., Robins, J., Douglas, S., Moody, C. and Cordingley, J. 2012. 'Cost-benefit analysis of ecological networks assessed through spatial analysis of ecosystem services'. *Journal of Applied Ecology*, 49: 571–80.
5. Vimal, R., Pluvinet, P., Sacca, C., Mazagol, P. O., Etlicher, B. and Thompson, J. D. 2012. 'Exploring spatial patterns of vulnerability for diverse biodiversity descriptors in regional conservation planning'. *Journal of Environmental Management*, 95: 9–16.
6. Wu, J., Wu, J., Wang, X. and Zhong, M. 2012. 'Securing water for wetland conservation: A comparative analysis of policy options to protect a national nature reserve in China'. *Journal of Environmental Management*, 94: 102–11.
7. Toledo-Aceves, T., Meave, J. A., González-Espinosa, M. and Ramírez-Marcial, N. 2011. 'Tropical montane cloud forests: Current threats and opportunities for their conservation and sustainable management in Mexico'. *Journal of Environmental Management*, 92: 974–81.

8. Fuller, D. O., Meijaard, E. M., Christy, L. and Jessup, T. C. 2010. 'Spatial assessment of threats to biodiversity within East Kalimantan, Indonesia'. *Applied Geography*, 30: 416–25.
9. Gorokhovich, Y. and Voustianiouk, A. 2010. 'Prioritization of coastal properties for conservation in New York State'. *Journal of Coastal Conservation*, 14: 41–51.
10. Valente, R. de O. A. and Vettorazzi, C. A. 2008. 'Definition of priority areas for forest conservation through the ordered weighted averaging method'. *Forest Ecology and Management*, 256: 1408–17.
11. Hajkowicz, S., Higgins, A., Miller, C. and Marinoni, O. 2008. 'Targeting conservation payments to achieve multiple outcomes'. *Biological Conservation*, 141: 2368–75.
12. Martínez-Harms, M. J. and Gajardo, R. 2008. 'Ecosystem value in the Western Patagonia protected areas'. *Journal for Nature Conservation*, 16: 72–87.
13. Dhar, A., Ruprecht, H. and Vacik, H. 2008. 'Population viability risk management (PVRM) for in situ management of endangered tree species: A case study on a Taxus baccata L. population'. *Forest Ecology and Management*, 255: 2835–45.
14. Regan, H. M., Davis, F. W., Andelman, S. J., Widyanata, A. and Freese, M. 2007. 'Comprehensive criteria for biodiversity evaluation in conservation planning'. *Biodiversity and Conservation*, 16: 2715–28.
15. Geneletti, D. 2007. 'An approach based on spatial multicriteria analysis to map the nature conservation value of agricultural land'. *Journal of Environmental Management*, 83: 228–35.
16. Ausseil, A. G. E., Dymond, J. R. and Shepherd, J. D. 2007. 'Rapid mapping and prioritisation of wetland sites in the Manawatu-Wanganui region, New Zealand'. *Environmental Management*, 39: 316–25.
17. Strager, M. P. and Rosenberger, R. S. 2006. 'Incorporating stakeholder preferences for land conservation: Weights and measures in spatial MCA'. *Ecological Economics*, 58: 79–92.
18. Moffett, A., Dyer, J. S. and Sarkar, S. 2006. 'Integrating biodiversity representation with multiple criteria in North-Central Namibia using non-dominated alternatives and a modified analytic hierarchy process'. *Biological Conservation*, 129: 181–91.
19. Geneletti, D. 2004. 'A GIS-based decision support system to identify nature conservation priorities in an alpine valley'. *Land Use Policy*, 21: 149–60.
20. Noss, R. F., Carroll, C., Vance-Borland, K. and Wuerthner, G.. 2002. 'A multicriteria assessment of the irreplaceability and vulnerability of sites in the Greater Yellowstone Ecosystem'. *Conservation Biology*, 16: 895–908.

21. Sierra, R., Campos, F. and Chamberlin, J. 2002. 'Assessing biodiversity conservation priorities: Ecosystem risk and representativeness in continental Ecuador'. *Landscape and Urban Planning*, 59: 95–110.
22. Farashi, A., Naderi, M. and Parvian, N. 2016. 'Identifying a preservation zone using multi-criteria decision analysis'. *Animal Biodiversity and Conservation*, 39: 29–36.
23. Habtemariam, B. T. and Fang, Q. 2016. 'Zoning for a multiple-use marine protected area using spatial multi-criteria analysis: The case of the Sheik Seid Marine National Park in Eritrea'. *Marine Policy*, 63: 135–43.
24. Etxano, I., Garmendia, E., Pascual, U., Hoyos, D., Diez, M.-A., Cadinanos, J. A. and Lozano, P. J. 2015. 'A participatory integrated assessment approach for Natura 2000 network sites'. *Environment and Planning C – Government and Policy*, 33: 1207–32.
25. Bali, A., Masoud Monavari, S., Riazi, B., Khorasani, N. and Masoud Kheirkhah Zarkesh, M. 2015. 'A spatial decision support system for ecotourism development in Caspian Hyrcanian Mixed Forests Ecoregion'. *Boletim de Ciências Geodésicas*, 21: 340–53.
26. Cortina, C. and Boggia, A. 2014. 'Development of policies for Natura 2000 sites: A multi-criteria approach to support decision makers'. *Journal of Environmental Management*, 141: 138–45.
27. Grazhdani, D. 2014. 'Integrating ecosystem services into assessment of different management options in a protected area: a deliberate multi-criteria decision analysis approach'. *Bulgarian Journal of Agricultural Science*, 20: 1311–19.
28. García-Marmolejo, G., Escalona-Segura, G. and Van Der Wal, H. 2008. 'Multicriteria evaluation of wildlife management units in Campeche, Mexico'. *Journal of Wildlife Management*, 72: 1194–202.
29. Lu, S. Y., Shen, C. H. and Chiau, W. Y. 2014. 'Zoning strategies for marine protected areas in Taiwan: Case study of Gucishan Island in Yilan County, Taiwan'. *Marine Policy*, 48: 21–29.
30. Zhang, Z., Sherman, R., Yang, Z., Wu, R., Wang, W., Yin, M., Yang, G. and Ou, X. 2013. 'Integrating a participatory process with a GIS-based multi-criteria decision analysis for protected area zoning in China'. *Journal for Nature Conservation*, 21: 225–40.
31. Hajehforooshnia, S., Soffianian, A., Mahiny, A. S. and Fakheran, S. 2011. 'Multi objective land allocation (MOLA) for zoning Ghamishloo Wildlife Sanctuary in Iran'. *Journal for Nature Conservation*, 19: 254–62.
32. Oikonomou, V., Dimitrakopoulos, P. G. and Troumbis, A. Y. 2011. 'Incorporating ecosystem function concept in environmental planning and decision making by means of multi-criteria evaluation: The case-study of Kalloni, Lesbos, Greece'. *Environmental Management*, 47: 77–92.

33. Zia, A., Hirsch, P., Songorwa, A., Mutekanga, D. R., O'Connor, S., McShane, T., Brosius, P. and Norton, B. 2011. 'Cross-scale value trade-offs in managing social-ecological systems: The politics of scale in Ruaha National Park, Tanzania'. *Ecology and Society*, 16(4): 7.
34. Khoi, D. D. and Murayama, Y. 2010. 'Delineation of suitable cropland areas using a GIS based multi-criteria evaluation approach in the tam Dao national park region, Vietnam.' *Sustainability*, 2: 2024–43.
35. Graziano, R., Gilberto, P. and Ferrarini, A. 2009. 'A rapid and cost-effective tool for managing habitats of the European Natura 2000 network: A case study in the Italian Alps'. *Biodiversity and Conservation*, 18: 1375–88.
36. Gurung, D. B. and Scholz, R. W. 2008. 'Community-based ecotourism in Bhutan: Expert evaluation of stakeholder-based scenarios'. *International Journal of Sustainable Development & World Ecology*, 15: 397–411.
37. Geneletti, D. and van Duren, I. 2008. 'Protected area zoning for conservation and use: A combination of spatial multicriteria and multiobjective evaluation'. *Landscape and Urban Planning*, 85: 97–110.
38. Wood, L. J. and Dragicevic, S. 2007. 'GIS-based multicriteria evaluation and fuzzy sets to identify priority sites for marine protection'. *Biodiversity and Conservation*, 16: 2539–58.
39. Boteva, D., Griffiths, G. and Dimopoulos, P. 2004. 'Evaluation and mapping of the conservation significance of habitats using GIS: An example from Crete, Greece'. *Journal for Nature Conservation*, 12: 237–50.
40. Grošelj, P., Hodges, D. G. and Zadnik Stirn, L. 2015. 'Participatory and multi-criteria analysis for forest (ecosystem) management: A case study of Pohorje, Slovenia'. *Forest Policy and Economics*, 71: 80–86.
41. Fernández, I. C. and Morales, N. S. 2016. 'A spatial multicriteria decision analysis for selecting priority sites for plant species restoration: a case study from the Chilean biodiversity hotspot'. *Restoration Ecology*, 24: 599–608.
42. Suryabhagavan, K. V, Alemu, M. and Balakrishnan, M. 2016. 'GIS-based multi-criteria decision analysis for forest fire susceptibility mapping: A case study in Harenna forest, southwestern Ethiopia'. *Tropical Ecology*, 57: 33–43.
43. Derak, M. and Cortina, J. 2014. 'Multi-criteria participative evaluation of Pinus halepensis plantations in a semiarid area of southeast Spain'. *Ecological Indicators*, 43: 56–68.
44. Lin, L., Sills, E. and Cheshire, H. 2014. 'Targeting areas for Reducing Emissions from Deforestation and forest Degradation (REDD+) projects in Tanzania'. *Global Environmental Change*, 24: 277–86.
45. Uribe, D., Geneletti, D., del Castillo, R. F. and Orsi, F. 2014. 'Integrating stakeholder preferences and GIS-based multicriteria analysis to

identify forest landscape restoration priorities'. *Sustainability (Switzerland)*, 6: 935–51.

46. Jactel, H., Branco, M., Duncker, P., Gardiner, B., Grodzki, W., Langstrom, B., Moreira, F., Netherer, S., Nicoll, B., Orazio, C., Piou, D., Schelhaas, M. J. and Tojic, K. 2012. 'A multicriteria risk analysis to evaluate impacts of forest management alternatives on forest health in Europe'. *Ecology and Society*, 17: 25.
47. Rantala, M., Hujala, T. and Kurttila, M. 2012. 'Measuring and monitoring socio-cultural sustainability in the action of forest biodiversity cooperation networks'. *Silva Fennica*, 46: 441–59.
48. Jalilova, G., Khadka, C. and Vacik, H. 2012. 'Developing criteria and indicators for evaluating sustainable forest management: A case study in Kyrgyzstan'. *Forest Policy and Economics*, 21: 32–43.
49. Schwenk, W. S., Donovan, T. M., Keeton, W. S. and Nunery, J. S. 2012. 'Carbon storage, timber production, and biodiversity: Comparing ecosystem services with multi-criteria decision analysis'. *Ecological Applications*, 22: 1612–27.
50. Nordström, E.-M., Eriksson, L. O. and Karin, Ö. 2011. 'Multiple criteria decision analysis with consideration to place-specific values in participatory forest planning'. *Silva fennica*, 45: 253–65.
51. Pasqualini, V., Oberti, P., Vigetta, S., Riffard, O., Panaïotis, C., Cannac, M. and Ferrat, L. 2011. 'A GIS-based multicriteria evaluation for aiding risk management Pinus pinaster Ait. Forests: A case study in Corsican island, western Mediterranean region'. *Environmental Management*, 48: 38–56.
52. Reubens, B., Moeremans, C., Poesen, J., Nyssen, J., Tewoldeberhan, S., Franzel, S., Deckers, J., Orwa, C. and Muys, B. 2011. 'Tree species selection for land rehabilitation in Ethiopia: From fragmented knowledge to an integrated multi-criteria decision approach'. *Agroforestry Systems*, 82: 303–30.
53. Mustajoki, J., Saarikoski, H., Marttunen, M., Ahtikoski, A., Hallikainen, V., Helle, T., Hyppönen, M., Jokinen, M., Naskali, A., Tuulentie, S., Varmola, M., Vatanen, E. and Ylisirniö, A. L. 2011. 'Use of decision analysis interviews to support the sustainable use of the forests in Finnish Upper Lapland'. *Journal of Environmental Management*, 92: 1550–63.
54. Ianni, E. and Geneletti, D. 2010. 'Applying the ecosystem approach to select priority areas for forest landscape restoration in the Yungas, northwestern Argentina'. *Environmental Management*, 46: 748–60.
55. Orsi, F. and Geneletti, D. 2010. 'Identifying priority areas for Forest Landscape Restoration in Chiapas (Mexico): An operational approach

combining ecological and socioeconomic criteria'. *Landscape and Urban Planning*, 94: 20–30.

56. Kint, V., Lasch, P., Lindner, M. and Muys, B. 2009. 'Multipurpose conversion management of Scots pine towards mixed oak-birch stands-A long-term simulation approach'. *Forest Ecology and Management*, 257: 199–214.
57. Briceño-Elizondo, E., Jäger, D., Lexer, M. J., Garcia-Gonzalo, J., Peltola, H. and Kellomäki, S. 2008. 'Multi-criteria evaluation of multi-purpose stand treatment programmes for Finnish boreal forests under changing climate'. *Ecological Indicators*, 8: 26–45.
58. Fürstenau, C., Badeck, F. W., Lasch, P., Lexer, M. J., Lindner, M., Mohr, P. and Suckow, F. 2007. 'Multiple-use forest management in consideration of climate change and the interests of stakeholder groups'. *European Journal of Forest Research*, 126: 225–39.
59. Van Der Horst, D. and Gimona, A. 2005. 'Where new farm woodlands support biodiversity action plans: A spatial multi-criteria analysis'. *Biological Conservation*, 123: 421–32.
60. Van Elegem, B., Elegem, B. V. A. N., Embo, T., Muys, B. and Lust, N. 2002. 'A methodology to select the best locations for new urban forests using multicriteria analysis'. *Forestry*, 75: 13–23.
61. Cordingley, J. E., Newton, A. C., Rose, R. J., Clarke, R. T. and Bullock, J. M. 2016. 'Can landscape-scale approaches to conservation management resolve biodiversity-ecosystem service trade-offs?' *Journal of Applied Ecology*, 53: 96–105.
62. Datta, D. and Kumar Ghosh, P. 2015. 'Evaluating sustainability of community endeavours in an Indian floodplain wetland using multi-criteria decision analysis'. *Singapore Journal of Tropical Geography*, 36: 38–56.
63. Hodder, K. H., Newton, A. C., Cantarello, E. and Perrella, L. 2014. 'Does landscape-scale conservation management enhance the provision of ecosystem services?' *International Journal of Biodiversity Science, Ecosystem Services & Management*, 10: 71–83.
64. Janssen, R., Knudsen, S., Todorova, V. and Hoşgör, A. G. 2014. 'Managing Rapana in the Black Sea: Stakeholder workshops on both sides'. *Ocean and Coastal Management*, 87: 75–87.
65. Forsyth, G. G., Le Maitre, D. C., O'Farrell, P. J. and van Wilgen, B. W. 2012. 'The prioritisation of invasive alien plant control projects using a multi-criteria decision model informed by stakeholder input and spatial data'. *Journal of Environmental Management*, 103: 51–57.
66. O'Connor, T. G., Kuyler, P., Kirkman, K. P. and Corcoran, B. 2010. 'Which grazing management practices are most appropriate for maintaining biodiversity in South African grassland?' *African Journal of Range & Forage Science*, 27: 67–76.

67. García de Jalón, S., Iglesias, A., Cunningham, R. and Pérez Díaz, J. I. 2014. 'Building resilience to water scarcity in southern Spain: A case study of rice farming in Doñana protected wetlands'. *Regional Environmental Change*, 14: 1229–42.
68. Oliver, I., Jones, H. and Schmoldt, D. L. 2007. 'Expert panel assessment of attributes for natural variability benchmarks for biodiversity'. *Austral Ecology*, 32: 453–75.
69. Rohde, S., Hostmann, M., Peter, A. and Ewald, K. C. 2006. 'Room for rivers: An integrative search strategy for floodplain restoration'. *Landscape and Urban Planning*, 78: 50–70.
70. Redpath, S. M., Arroyo, B. E., Leckie, F. M., Bacon, P., Bayfield, N., Guttiérrez, R. J. and Thirgood, S. J. 2004. 'Using decision modeling with stakeholders to reduce human-wildlife conflict: A raptor-grouse case study'. *Conservation Biology*, 18: 350–59.
71. Riccioli, F., Fratini, R., Boncinelli, F., El Asmar, T., El Asmar, J. P. and Casini, L. 2016. 'Spatial analysis of selected biodiversity features in protected areas: A case study in Tuscany region'. *Land Use Policy*, 57: 540–54.
72. Gülci, S. and Akay, A. E. 2015. 'Assessment of ecological passages along road networks within the Mediterranean forest using GIS-based multi criteria evaluation approach'. *Environmental Monitoring and Assessment*, 187: 1–13.
73. Comino, E., Bottero, M., Pomarico, S. and Rosso, M. 2014. 'Exploring the environmental value of ecosystem services for a river basin through a spatial multicriteria analysis'. *Land Use Policy*, 36: 381–95.
74. Orsi, F., Geneletti, D. and Borsdorf, A. 2013. 'Mapping wildness for protected area management: A methodological approach and application to the Dolomites UNESCO World Heritage Site (Italy)'. *Landscape and Urban Planning*, 120: 1–15.
75. Momeni Dehaghi, I., Mahiny, A. S., Shabani, A. A., Karami, M., Taylor, P. and Dehaghi, I. M. 2013. 'Efficiency of current reserve network in Golestan Province (Iran) for the protection of hoofed ungulates'. *Biodiversity*, 14: 162–68.
76. Carver, S., Comber, A., McMorran, R. and Nutter, S. 2012. 'A GIS model for mapping spatial patterns and distribution of wild land in Scotland'. *Landscape and Urban Planning*, 104: 395–409.
77. Geneletti, D. 2008. 'Incorporating biodiversity assets in spatial planning: Methodological proposal and development of a planning support system'. *Landscape and Urban Planning*, 84: 252–65.
78. Karlson, M., Karlsson, C. S. J., Mörtberg, U., Olofsson, B. and Balfors, B. 2016. 'Design and evaluation of railway corridors based on spatial

ecological and geological criteria'. *Transportation Research Part D: Transport and Environment*, 46: 207–28.

79. Marre, J. B., Pascoe, S., Thébaud, O., Jennings, S., Boncoeur, J. and Coglan, L. 2016. 'Information preferences for the evaluation of coastal development impacts on ecosystem services: A multi-criteria assessment in the Australian context'. *Journal of Environmental Management*, 173: 141–50.
80. Ferretti, V. and Comino, E. 2015. 'An integrated framework to assess complex cultural and natural heritage systems with Multi-Attribute Value Theory'. *Journal of Cultural Heritage*, 16: 688–97.
81. Stoms, D. M., Dashiell, S. L. and Davis, F. W. 2013. 'Siting solar energy development to minimize biological impacts'. *Renewable Energy*, 57: 289–98.
82. Alavi, I. and Alinejad-Rokny, H. 2011. 'Comparison of Fuzzy AHP and Fuzzy TOPSIS methods for plant species selection (case study: Reclamation plan of Sungun Copper Mine; Iran)'. *Australian Journal of Basic and Applied Sciences*, 5: 1104–13.
83. Reig, E., Aznar, J. and Estruch, V. 2010. 'A comparative analysis of the sustainability of rice cultivation technologies using the analytic network process'. *Spanish Journal of Agricultural Research*, 8: 273–84.
84. Snep, R., Van Ierland, E. and Opdam, P. 2009. 'Enhancing biodiversity at business sites: What are the options, and which of these do stakeholders prefer?' *Landscape and Urban Planning*, 91: 26–35.
85. Svoray, T., Bar, P. and Bannet, T. 2005. 'Urban land-use allocation in a Mediterranean ecotone: Habitat Heterogeneity Model incorporated in a GIS using a multi-criteria mechanism'. *Landscape and Urban Planning*, 72: 337–51.
86. Curtis, I. A. 2004. 'Valuing ecosystem goods and services: A new approach using a surrogate market and the combination of a multiple criteria analysis and a Delphi panel to assign weights to the attributes. *Ecological Economics*, 50: 163–94.

REFERENCES

Adem Esmail, Blal, and Davide Geneletti. 2018. 'Multi-criteria Decision Analysis for Nature Conservation: A Review of 20 Years of Applications'. *Methods in Ecology and Evolution* 9 (1): 42–53. doi:10.1111/2041-210X.12899.

Allison, W. R. 1996. 'Snorkeler Damage to Reef Corals in the Maldive Islands'. *Coral Reefs* 15 (4): 215–18. doi:10.1007/BF01787454.

Amo, Luisa, Pilar López and José Martín. 2007. 'Habitat Deterioration Affects Body Condition of Lizards: A Behavioral Approach with Iberolacerta Cyreni Lizards Inhabiting Ski Resorts'. *Biological Conservation* 135 (1): 77–85. doi:10.1016/j.biocon.2006.09.020.

Arnaud-Fassetta, Gilles, Etienne Cossart and Monique Fort. 2005. 'Hydro-Geomorphic Hazards and Impact of Man-Made Structures during the Catastrophic Flood of June 2000 in the Upper Guil Catchment (Queyras, Southern French Alps)'. *Geomorphology* 66 (1–4 SPEC. ISS.): 41–67. doi:10.1016/j.geomorph.2004.03.014.

Bali, A., S. Masoud Monavari, B. Riazi, N. Khorasani and M. Masoud Kheirkhah Zarkesh. 2015. 'A Spatial Decision Support System for Ecotourism Development in Caspian Hyrcanian Mixed Forests Ecoregion'. *Boletim de Ciências Geodésicas* 21: 340–53.

Beierle, Thomas C. 2002. 'The Quality of Stakeholder-Based Decisions'. *Risk Analysis* 22 (4): 739–49.

Beinat, Euro. 1997. *Value Functions for Environmental Management.* Dordrecht: Springer Netherlands.

Beinat, Euro, and Peter Nijkamp. 1998. *Multicriteria Analysis for Land-Use Management.* Vol. 9. Dordrecht: Kluwer Academic Publishers. doi:10.1007/978-94-015-9058-7.

Belton, Valerie, and Theo Stewart. 2002. *Multiple Criteria Decision Analysis: An Integrated Approach.* Boston: Kluwer Academic Publishers.

Bhushan, N., and K. Rai. 2007. *Strategic Decision Making: Applying the Analytic Hierarchy Process.* Springer Science & Business Media.

Bogaert, J., R. Rousseau, P. Van Hecke and I. Impens. 2000. 'Alternative Area-Perimeter Ratios for Measurement of 2D Shape Compactness of Habitats'. *Applied Mathematics and Computation* 111 (1): 71–85. doi:10.1016/S0096-3003(99)00075-2.

Bojórquez-Tapia, Luis A., Horacio De La Cueva, Salomón Díaz, Daniela Melgarejo, Georgina Alcantar, María José Solares, Gerardo Grobet and Gustavo Cruz-Bello. 2004. 'Environmental Conflicts and Nature Reserves: Redesigning Sierra San Pedro Mártir National Park, Mexico'. *Biological Conservation* 117 (2): 111–26. doi:10.1016/S0006-3207(03)00265-9.

Breedlove, Dennis E. 1981. 'Flora of Chiapas. Part 1: Introduction to the Flora of Chiapas'. San Francisco: California Academy of Science.

Burrough, Peter A., and Rachael A. McDonnell. 1998. *Principles of Geographical Information Systems*. Oxford: Oxford University Press.

Canter, Larry W. 1996. *Environmental Impact Assessment*. 2nd ed. New York: McGraw-Hill.

Cavallin, A., M. Marchetti and M. Panizza. 1996. 'Geomorphology and Environmental Impact Assessment: A Methodological Approach'. In M. Panizza, A. G. Fabbri, M. Marchetti and A. Patrono (eds), *Geomorphologic Analysis and Evaluation in Environmental Impact Assessment*, 308–10. Enschede: ITC Publication 32.

Cegan, Jeffrey C., Ashley M. Filion, Jeffrey M. Keisler and Igor Linkov. 2017. 'Trends and Applications of Multi-criteria Decision Analysis in Environmental Sciences: Literature Review'. *Environment Systems and Decisions* 37 (2). Springer US: 123–33. doi:10.1007/s10669-017-9642-9.

Choo, Eng U., Bertram Schoner and William C. Wedley. 1999. 'Interpretation of Criteria Weights in Multicriteria Decision Making'. *Computers and Industrial Engineering* 37 (3): 527–41. doi:10.1016/S0360-8352(00)00019-X.

Comino, E., M. Bottero, S. Pomarico and M. Rosso . 2014. 'Exploring the Environmental Value of Ecosystem Services for a River Basin through a Spatial Multicriteria Analysis'. *Land Use Policy* 36: 381–95.

CONAPO (Consejo Nacional de Poblacion). 2005. 'Índices de Marginación, 2005 (Marginalisation Indices, 2005)'. Mexico, DF.

Costa, Carlos A Bana e and Rui Carvalho Oliveira. 2002. 'Assigning Priorities for Maintenance, Repair and Refurbishment in Managing a Municipal Housing Stock'. *European Journal of Operational Research* 138 (2): 380–91.

Creachbaum, M. S., C. Johnson and R. H. Schmidt. 1998. 'Living on the Edge: A Process for Redesigning Campgrounds in Grizzly Bear Habitat'. *Landscape and Urban Planning* 42 (2–4): 269–86. doi:10.1016/S0169-2046(98)00092-9.

Crossman, N. D., B. Ostendorf, B. A. Bryan, A. Nefiodovas and A. Wright. 2005. 'OSS: A Spatial Decision Support System for Optimal Zoning of Marine Protected Areas'. In *Modsim 2005: International Congress on Modelling and Simulation: Advances and Applications for Management and Decision Making*, 1525–31.

DCLG. (2009). Multi-criteria Analysis: A Manual. West Yorkshire: Department for Communities and Local Government (DCLG).

Delgado, Montserrat Gómez, and Joaquín Bosque Sendra. 2004. 'Sensitivity Analysis in Multicriteria Spatial Decision-Making: A Review'. *Human and Ecological Risk Assessment: An International Journal* 10 (6).

Delgado, Otoniel Buenrostro, Manuel Mendoza, Erna López Granados and Davide Geneletti. 2008. 'Analysis of Land Suitability for the Siting of Inter-municipal Landfills in the Cuitzeo Lake Basin, Mexico'. *Waste Management* 28 (7): 1137–46. doi:10.1016/j.wasman.2007.07.002.

Derak, McHich, and Jordi Cortina. 2014. 'Multi-criteria Participative Evaluation of Pinus Halepensis Plantations in a Semiarid Area of Southeast Spain'. *Ecological Indicators* 43. Elsevier Ltd: 56–68. doi:10.1016/j.ecolind.2014.02.017.

Dodgson, J. S., M. Spackman, A. Pearman and L. D. Phillips. 2009. 'Multi-criteria Analysis : A Manual'. *Department for Communities and Local Government*. Vol. 11. London. doi:10.1002/mcda.399.

Eastman, J. Ronald. 2003. 'IDRISI Kilimanjaro Guide to GIS and Image Processing'. Worcester, MA: ClarkLabs.

Eastman, J. Ronald, Hong Jiang and James Toledano. 1998. 'Multi-criteria and Multi-objective Decision Making for Land Allocation Using GIS'. In E. Beinat and P. Nijkamp (eds), *Multicriteria Analysis for Land-Use Management*, 227–51. Dordrecht: Kluwer Academic Publishers.

Eastman, Ronald, Weigen Jin, Peter A. K. Kyem and James Toledano. 1995. 'For Procedures Raster Decisions M Ulti-Criteria'. *Photogrammetric Engineering and Remote Sensing* 61 (5): 539–47.

Elegem, B. Van, T. Embo, B. Muys and N. Lust. 2002. 'A Methodology to Select the Best Locations for New Urban Forests Using Multicriteria Analysis'. *Forestry* 75 (1): 13–23. doi:10.1093/forestry/75.1.13.

Environmental Protection Authority (EPA). 2007. 'Guidelines for Separation Distances'. Adelaide.

European Commission. 2003. 'Common Implementation Strategy for the Water Framework Directive (2000/60/CE) – Guidance Document n. 8'.

———. 2006. 'Development of a Methodology to Assess Population Exposed to High Levels of Noise and Air Pollution Close to Major Transport Infrastructure'. London.

Ferretti, Valentina, and Elena Comino. 2015. 'An Integrated Framework to Assess Complex Cultural and Natural Heritage Systems with Multi-Attribute Value Theory'. *Journal of Cultural Heritage* 16 (5). Elsevier Masson SAS: 688–97. doi:10.1016/j.culher.2015.01.007.

Ferretti, Valentina, and Silvia Pomarico. 2013. 'An Integrated Approach for Studying the Land Suitability for Ecological Corridors through Spatial Multicriteria Evaluations'. *Environment, Development and Sustainability* 15 (3): 859–85. doi:10.1007/s10668-012-9400-6.

French, Simon, John Maule and Nadia Papamichail. 2009. *Decision Behaviour, Analysis and Support.* 1st ed. New York: Cambridge University Press.

Gamper, C. D., and C. Turcanu. 2007. 'On the Governmental Use of Multi-criteria Analysis'. *Ecological Economics* 62 (2): 298–307. doi:10.1016/j.ecolecon.2007.01.010.

García-Marmolejo, Gabriela, Griselda Escalona-Segura and Hans Van Der Wal. 2008. 'Multicriteria Evaluation of Wildlife Management Units in Campeche, Mexico'. *Journal of Wildlife Management* 72 (5): 1194–202. doi:10.2193/2006–050.

Geneletti, Davide. 2004a. 'Using Spatial Indicators and Value Functions to Assess Ecosystem Fragmentation Caused by Linear Infrastructures'. *International Journal of Applied Earth Observation and Geoinformation* 5 (1): 1–15. doi:10.1016/j.jag.2003.08.004.

———. 2004b. 'A GIS-Based Decision Support System to Identify Nature Conservation Priorities in an Alpine Valley'. *Land Use Policy* 21 (2): 149–60. doi:10.1016/j.landusepol.2003.09.005.

———. 2005a. 'Formalising Expert Opinion through Multi-attribute Value Functions: An Application in Landscape Ecology'. *Journal of Environmental Management* 76 (3): 255 62.

———. 2005b. 'Multicriteria Analysis to Compare the Impact of Alternative Road Corridors: A Case Study in Northern Italy'. *Impact Assessment and Project Appraisal* 23 (2): 135–46. doi:10.3152/147154605781765661.

———. 2007a. 'Expert Panel-Based Assessment of Forest Landscapes for Land Use Planning'. *Mount* 27 (3): 220–23. doi:10.1659/mrd.0924.

———. 2007b. 'An Approach Based on Spatial Multicriteria Analysis to Map the Nature Conservation Value of Agricultural Land'. *Journal of Environmental Management* 83 (2): 228–35. doi:10.1016/j.jenvman.2006.03.002.

———. 2008a. 'Incorporating Biodiversity Assets in Spatial Planning: Methodological Proposal and Development of a Planning Support System'. *Landscape and Urban Planning* 84 (3–4): 252–65. doi:10.1016/j.landurbplan.2007.08.005.

———. 2008b. 'Impact Assessment of Proposed Ski Areas: A GIS Approach Integrating Biological, Physical and Landscape Indicators'. *Environmental Impact Assessment Review* 28 (2–3): 116–30. doi:10.1016/j.eiar.2007.05.011.

———. 2010. 'Combining Stakeholder Analysis and Spatial Multicriteria Evaluation to Select and Rank Inert Landfill Sites'. *Waste Management* 30 (2): 328–37.

———. 2013. 'Assessing the Impact of Alternative Land-Use Zoning Policies on Future Ecosystem Services'. *Environmental Impact Assessment Review* 40 (2013): 25–35.

———. 2014. 'Integration of Impact Assessment Types Improves Consideration of Alternatives'. *Impact Assessment and Project Appraisal* 32(1): 17–18. doi:10.1080/14615517.2013.872846.

Geneletti, Davide, and Iris van Duren. 2008. 'Protected Area Zoning for Conservation and Use: A Combination of Spatial Multicriteria and Multiobjective Evaluation'. *Landscape and Urban Planning* 85 (2): 97–110. doi:10.1016/j.landurbplan.2007.10.004.

Geneletti, Davide, and Valentina Ferretti. 2015. 'Multicriteria Analysis for Sustainability Assessment: Concepts and Case Studies'. In *Handbook of Sustainability Assessment.* Cheltenham, UK; Northampton, MA: Edward Elgar Publishing.

George, Annette. 2003. 'Managing Ski Resorts: Perceptions from the Field Regarding the Sustainable Slopes Charter'. *Managing Leisure* 8 (1): 41–46.

Gibson, R. B., S. Hassan, S. Holtz, J. Tansey and G. Whitelaw. 2005. 'Sustainability Assessment Criteria, Processes and Applications'. *Earthscan, London.*

Gonzalez, Ainhoa, and Álvaro Enríquez-de-Salamanca. 2018. 'Spatial Multi-criteria Analysis in Environmental Assessment: A Review and Reflection on Benefits and Limitations'. *Journal of Environmental Assessment Policy and Management* 20.03 (2018): 1840001.

Gregory, Robin, Lee Failing and Michael Harstone. 2012. *Structured Decision Making: A Practical Guide to Environmental Management Choices.* Chichester: Wiley-Blackwell.

Grimble, Robin, and Kate Wellard. 1997. 'Stakeholder Methodologies in Natural Resource Management: A Review of Principles, Contexts, Experiences and Opportunities'. *Agricultural Systems* 55 (2): 173–93. doi:10.1016/S0308-521X(97)00006-1.

Gros, Raphaël, Lucile Jocteur Monrozier, François Bartoli, Jean Luc Chotte and Pierre Faivre. 2004. 'Relationships between Soil Physico-chemical Properties and Microbial Activity along a Restoration Chronosequence of Alpine Grasslands Following Ski Run Construction'. *Applied Soil Ecology* 27 (1): 7–22. doi:10.1016/j.apsoil.2004.03.004.

Gustafson, Eric. 1998. 'Pattern : What Is the State of the Art ?' *Ecosystems* I: 143–56. doi:10.1007/s100219900011.

Habtemariam, Bereket Tesfamariam and Qinhua Fang. 2016. 'Zoning for a Multiple-Use Marine Protected Area Using Spatial Multi-criteria Analysis: The Case of the Sheik Seid Marine National Park in Eritrea'. *Marine Policy* 63. Elsevier: 135–43. doi:10.1016/j.marpol.2015.10.011.

Hadley, Gillian L., and Kenneth R. Wilson. 2004. 'Patterns of Density and Survival in Small Mammals in Ski Runs and Adjacent Forest Patches'. *The Journal of Wildlife Management* 68 (2): 288–98.

Herwijnen, M. van. 1999. *Spatial Decision Support for Environmental Management.* Amsterdam: Tinbergen Instituut.

Hess, George R., and Terri J. King. 2002. 'Planning Open Spaces for Wildlife I. Selecting Focal Species Using a Delphi Survey Approach'. *Landscape and Urban Planning* 58 (1): 25–40. doi:10.1016/S0169-2046(01)00230-4.

Higgs, Gary. 2006. 'Integrating Multi-criteria Techniques with Geographical Information Systems in Waste Facility Location to Enhance Public Participation'. *Waste Management and Research* 24 (2): 105–17. doi:10.1177/0734242X06063817.

Hirzel, Alexandre H., Gwenaëlle Le Lay, Véronique Helfer, Christophe Randin and Antoine Guisan. 2006. 'Evaluating the Ability of Habitat Suitability Models to Predict Species Presences'. *Ecological Modelling* 199 (2): 142–52. doi:10.1016/j.ecolmodel.2006.05.017.

Hjortsø, Carsten Nico, Steffen Stræde and Finn Helles. 2006. 'Applying Multi-criteria Decision-Making to Protected Areas and Buffer Zone Management: A Case Study in the Royal Chitwan National Park, Nepal'. *Journal of Forest Economics* 12 (2): 91–108. doi:10.1016/j.jfe.2006.03.001.

Hobbs, R. J. 1993. 'Effects of Landscape Fragmentation on Ecosystem Processes in the Western Australian Wheatbelt'. *Biological Conservation* 64 (3): 193–201. doi:10.1016/0006-3207(93)90321-Q.

Hobbs, R. J., and J. A. Harris. 2001. 'Restoration Ecology : Repairing the Earth's Ecosystems in the New Millennium'. *Restoration Ecology* 9 (2): 239–46. doi:10.1046/j.1526-100x.2001.009002239.x.

Hokkanen, J., and P. Salminen. 1997. 'Locating a Waste Treatment Facility by Multicriteria Analysis'. *Journal of Multi-Criteria Decision Analysis* 6 (3): 175–84.

Hostovsky, C. 2000. 'Integrating Planning Theory and Waste Management: An Annotated Bibliography'. *Journal of Planning Literature* 15: 305–32.

Huang, Ivy B., Jeffrey Keisler and Igor Linkov. 2011. 'Multi-criteria Decision Analysis in Environmental Sciences: Ten Years of Applications and Trends'. *Science of the Total Environment* 409 (19). Elsevier B.V.: 3578–94. doi:10.1016/j.scitotenv.2011.06.022.

Huby, Meg, Anne Owen and Steve Cinderby. 2007. 'Reconciling Socio-economic and Environmental Data in a GIS Context: An Example from Rural England'. *Applied Geography* 27 (1): 1–13. doi:10.1016/j.apgeog.2006.10.001.

Hwang, Ching-Lai, and Kwangsun Yoon. 1981. *Multiple Attribute Decision Making: Methods and Applications A State-of-the-Art Survey*. Berlin Heidelberg: Springer-Verlag.

Ianni, Elena, and Davide Geneletti. 2010. 'Applying the Ecosystem Approach to Select Priority Areas for Forest Landscape Restoration in the Yungas, Northwestern Argentina'. *Environmental Management* 46 (5): 748–60. doi:10.1007/s00267-010-9553-8.

International Tropical Timber Organization (ITTO), and International Union for Conservation of Nature (IUCN). 2005. 'Restoring Forest Landscapes'. Yokohama, Japan.

Jalilova, G., C. Khadka and H. Vacik. 2012. 'Developing Criteria and Indicators for Evaluating Sustainable Forest Management: A Case Study in Kyrgyzstan'. *Forest Policy and Economics* 21: 32–43.

Jamieson, Bruce, and Chris Stethem. 2002. 'Snow Avalanche Hazards and Management in Canada: Challenges and Progress'. *Natural Hazards* 26 (1): 35–53. doi:10.1023/A:1015212626232.

Janssen, R., V. Herwijnen and E. Beinat. 2001. 'DEFINITE. Case Studies and User Manual. Report R-01/01'. Amsterdam.

Janssen, Ron. 2001. 'On the Use of Multi-criteria Analysis in Environmental Impact Assessment in the Netherlands'. *Journal of Multi-Criteria Decision Analysis* 10 (2): 101–9. doi:10.1002/mcda.293.

Janssen, Ron, Marjan van Herwijnen and Euro Beinat. 2003. 'DEFINITE 3.0 Case Studies and User's Manual'. Amsterdam: Institute for Environmental Studies, Vrije Universiteit.

Janssen, Ron, Ståle Knudsen, Valentina Todorova and Ayşe Gündüz Hoşgör. 2014. 'Managing Rapana in the Black Sea: Stakeholder Workshops on Both Sides'. *Ocean and Coastal Management* 87: 75–87. doi:10.1016/j.ocecoaman.2013.10.015.

Jongman, Rob H. G. 1995. 'Nature Conservation Planning in Europe: Developing Ecological Networks'. *Landscape and Urban Planning* 32 (3): 169–83. doi:10.1016/0169-2046(95)00197-O.

Kangas, Jyrki, and Pekka Leskinen. 2005. 'Modelling Ecological Expertise for Forest Planning Calculations-Rationale, Examples, and Pitfalls'. *Journal of Environmental Management* 76 (2): 125–33. doi:10.1016/j.jenvman.2005.01.011.

Karlson, M., C. S. J. Karlsson, U. Mörtberg, B. Olofsson and B. Balfors. 2016. 'Design and Evaluation of Railway Corridors Based on Spatial Ecological and Geological Criteria'. *Transportation Research Part D: Transport and Environment* 46. Elsevier Ltd: 207–28. doi:10.1016/j.trd.2016.03.012.

Keeney, R. L., and T. L. McDaniels. 1992. 'Value-Focused Thinking about Strategic Decisions at BC Hydro'. *Interfaces* 22 (6): 94–109. doi:10.1287/inte.22.6.94.

Keeney, Ralph L. 1992. *Value-Focused Thinking: A Path to Creative Decision Making.* Cambridge, MA: Harvard University Press.

———. 2002. 'Commom Mistakes in Making Value Trade-Offs.Pdf'. *Operations Research 2* 50 (6): 935–45.

Keeney, Ralph L., and Howard Raiffa. 1976. *Decisions with Multiple Objectives: Preferences and Value Tradeoffs.* New York: John Wiley & Sons.

Keisler, Jeffrey, and Ronald Sundell. 1997. 'Combining Multi-attribute Utility and Geographic Information for Boundary Decisions: An Application to Park Planning'. *Journal of Geographic Information and Decision Analysis* 1 (2): 100–19. http://scholarworks.umb.edu/msis_faculty_pubs%5Cnhttp://scholarworks.umb.edu/msis_faculty_pubs/36.

Kiker, Gregory A., Todd S. Bridges, Arun Varghese, Thomas P. Seager and Igor Linkov. 2005. 'Application of Multicriteria Decision Analysis in Environmental Decision Making'. *Integrated Environmental Assessment and Management* 1 (2): 95–108. doi:10.1897/IEAM_2004a-015.1.

Kirkpatrick, J. 2004. 'Landscape Values'. In National Parks & Wildlife Service (ed.), *An Assessment of the Values of Kosciuszko National Park.* NSW: National Parks & Wildlife Service.

Kliskey, A. D. 2000. 'Recreation Terrain Suitability Mapping: A Spatially Explicit Methodology for Determining Recreation Potential for Resource Use Assessment.' *Landscape and Urban Planning* 52 (1): 33–43. doi:10.1016/S0169-2046(00)00111-0.

Kurth, Margaret H., Sabrina Larkin, Jeffrey M. Keisler and Igor Linkov. 2017. 'Trends and Applications of Multi-criteria Decision Analysis: Use in Government Agencies'. *Environment Systems and Decisions* 37 (2). Springer US: 134–43. doi:10.1007/s10669-017-9644-7.

Laiolo, Paola, and Antonio Rolando. 2005. 'Forest Bird Diversity and Ski-Runs: A Case of Negative Edge Effect'. *Animal Conservation* 8 (1): 9–16. doi:10.1017/S1367943004001611.

Lamb, David, Peter D. Erskine, John A. Parotta and Forest Landscapes. 2005. 'Restoration of Degraded Tropical Forest Landscapes'. *Science* 310 (December): 1628–32.

Linkov, I., F. K. Satterstrom, G. Kiker, T. P. Seager, T. Bridges, K. H. Gardner, S. H. Rogers, D. A. Belluck and A. Meyer. 2006. 'Multicriteria Decision Analysis: A Comprehensive Decision Approach for Management of Contaminated Sediments'. *Risk Analysis* 26 (1): 61–78. doi:10.1111/j.1539-6924.2006.00713.x.

Linkov, Igor, and Emily Moberg. 2012. *Multi-criteria Decision Analysis: Environmental Applications and Case Studies.* United States: CRC Press.

Lu, Shiau Yun, Cheng Han Shen and Wen Yan Chiau. 2014. 'Zoning Strategies for Marine Protected Areas in Taiwan: Case Study of Gueishan Island in Yilan County, Taiwan'. *Marine Policy* 48. Elsevier: 21–29. doi:10.1016/j.marpol.2014.03.001.

MacMillan, Douglas C., and K. Marshall. 2006. 'The Delphi Process – An Expert-Based Approach to Ecological Modelling in Data-Poor Environments'. *Animal Conservation* 9 (1): 11–19. doi:10.1111/j.1469-1795.2005.00001.x.

Malczewski, Jacek. 1999. *GIS and Multicriteria Decision Analysis*. New York: John Wiley & Sons.

———. 2004. 'GIS-Based Land-Use Suitability Analysis: A Critical Overview'. *Progress in Planning* 62 (1): 3–65. doi:10.1016/j.progress.2003.09.002.

Marre, Jean Baptiste, Sean Pascoe, Olivier Thébaud, Sarah Jennings, Jean Boncoeur and Louisa Coglan. 2016. 'Information Preferences for the Evaluation of Coastal Development Impacts on Ecosystem Services: A Multi-criteria Assessment in the Australian Context'. *Journal of Environmental Management* 173: 141–50. doi:10.1016/j.jenvman.2016.01.025.

McDougall, K. L., J. W Morgan, N. G. Walsh and R. J. Williams. 2005. 'Plant Invasions in Treeless Vegetation of the Australian Alps'. *Perspectives in Plant Ecology, Evolution and Systematics* 7 (3): 159–71.

McGarigal, Kevin, and Barbara J. Marks. 1995. 'FRAGSTATS: Spatial Pattern Analysis Program for Quantifying Landscape Structure'. US Department of Agriculture, Forest Service, Pacific Northwest Research Station.

Miller, James R., and Richard J. Hobbs. 2007. 'Habitat Restoration? Do We Know What We Are Doing?' *Restoration Ecology* 15 (3): 382–90. doi:10.1111/j.1526-100X.2007.00234.x.

Moleele, N. M., and J. Mainah. 2003. 'Resource Use Conflicts: The Future of the Kalahari Ecosystem'. *Journal of Arid Environments* 54 (2): 405–23. doi:10.1006/jare.2002.1099.

Mosimann, Thomas. 1985. 'Geo-ecological Impacts of Ski Piste Construction in the Swiss Alps'. *Applied Geography* 5 (1): 29–37. doi:10.1016/0143-6228(85)90004-9.

Mourmouris, John C. 2006. 'MCDM for Location and Evaluation of Waste Disposal Facilities in Metropolitan Areas: The Case of East Attica'. *International Journal of Environmental Technology and Management* 6 (1–2): 180–92.

Mundaa, G., P. Nijkamp and P. Rietveld. 1995. 'Qualitative Multicriteria Methods for Fuzzy Evaluation Problems: An Illustration of Economic-Ecological Evaluation'. *European Journal of Operational Research* 82 (1): 79–97.

Mustajoki, Jyri, and Mika Marttunen. 2017. 'Comparison of Multi-criteria Decision Analytical Software for Supporting Environmental Planning Processes'. *Environmental Modelling and Software* 93. Elsevier Ltd: 78–91. doi:10.1016/j.envsoft.2017.02.026.

Mustajoki, Jyri, Heli Saarikoski, Mika Marttunen, Anssi Ahtikoski, Ville Hallikainen, Timo Helle, Mikko Hyppönen et al. 2011. 'Use of Decision Analysis Interviews to Support the Sustainable Use of the Forests in Finnish Upper Lapland'. *Journal of Environmental Management* 92 (6). Elsevier Ltd: 1550–63. doi:10.1016/j.jenvman.2011.01.007.

Myers, Norman, Russell A. Mittermeier, Cristina G. Mittermeier, Gustavo A. B. da Fonseca and Jennifer Kent. 2000. 'Biodiversity Hotspots for Conservation Priorities'. *Nature* 403 (6772): 853–58. doi:10.1038/35002501.

Naveh, Z. 1994. 'From Biodiversity to Ecodiversity: A Landscape-Ecology Approach to Conservation and Restoration'. *Restoration Ecology*. doi:10.1111/j.1526-100X.1994.tb00065.x.

Newton, A., and V. Kapos. 2003. 'Restoration of Wooded Landscapes: Placing UK Initiatives in a Global Context'. In J. Humphrey, A. Newton, J. Latham, H. Gray, K. Kirby, E. Poulsom et al., *The Restoration of Wooded Landscapes*, 7–24. Edinburgh: Forestry Commission.

Nordström, Eva-Maria, Ljusk Ola Eriksson and Öhman Karin. 2011. 'Multiple Criteria Decision Analysis with Consideration to Place-Specific Values in Participatory Forest Planning'.

Norese, Maria Franca. 2006. 'ELECTRE III as a Support for Participatory Decision-Making on the Localisation of Waste-Treatment Plants'. *Land Use Policy* 23 (1): 76–85. doi:10.1016/j.landusepol.2004.08.009.

Noss, R. F., and A. Y. Cooperrider. 1994. *Saving Nature's Legacy: Protecting and Restoring Biodiversity*. Washington DC: Island Press.

Noss, Reed F. 2000. 'High-Risk Ecosystems as Foci for Considering Biodiversity and Ecological Integrity in Ecological Risk Assessments'. *Environmental Science and Policy* 3 (6): 321–32. doi:10.1016/S1462-9011(00)00112-X.

OECD. 2006. 'Applying Strategic Environmental Assessment: Good Practice Guidance for Development Co-operation'. DAC Guidelines and Reference Series. Organisation for Economic Cooperation and Development.

Oliver, Ian. 2002. 'An Expert Panel-Based Approach to the Assessment of Vegetation Condition within the Context of Biodiversity Conservation Stage 1: The Identification of Condition Indicators'. *Ecological Indicators* 2 (3): 223–37. doi:10.1016/S1470-160X(02)00025-0.

Orsi, F., D. Geneletti and A. C. Newton. 2011. 'Towards a Common Set of Criteria and Indicators to Identify Forest Restoration Priorities: An Expert Panel-Based Approach'. *Ecological Indicators* 2 (11): 337–47.

Orsi, Francesco, and Davide Geneletti. 2010. 'Identifying Priority Areas for Forest Landscape Restoration in Chiapas (Mexico): An Operational Approach Combining Ecological and Socioeconomic Criteria'. *Landscape and Urban Planning* 94 (1): 20–30. doi:10.1016/j.landurbplan.2009.07.014.

Petts, J. 2001. 'Evaluating the Effectiveness of Deliberative Processes: Waste Management Case-Studies'. *Journal of Environmental Planning and Management* 44 (2): 207–26. doi:10.1080/09640560120033713.

Pfeffer, K. 2003. *Integrating Spatio-temporal Environmental Models for Planning Ski Runs. Nederlandse Geografische Studies*. Utrecht: Netherlands Geographical Studies.

Prosser, Filippo. 2001. *Lista Rossa Della Flora Del Trentino. Pteridofite e Fanerogame*. Rovereto, Osiride: Museo Civico di Rovereto.

R Development Core Team. 2005. 'R: A Language and Environment for Statistical Computing'. Vienna: R foundation for Statistical Computing.

Ramsar. 2006. 'Strategic Framework for the List of Wetlands of International Importance'. doi:http://www.ramsar.org/key_guide_list2006_e.pdf.

Renard, K., F. Weesies, D. McCool and D. Yoder. 1997. 'Prediction of Soil Erosion by Water: A Guide to Conservation Planning with the Revised Universal Soil Loss Equations (RUSLE)'. Agric. Handbook 703. Washington DC: USDA.

Retief, Francois, Angus Morrison-Saunders, Davide Geneletti and Jenny Pope. 2013. 'Exploring the Psychology of Trade-off Decision-Making in Environmental Impact Assessment'. *Impact Assessment and Project Appraisal* 31 (1): 13–23. doi:10.1080/14615517.2013.768007.

Reubens, Bert, Clara Moeremans, Jean Poesen, Jan Nyssen, Sarah Tewoldeberhan, Steve Franzel, Jozef Deckers, Caleb Orwa and Bart Muys. 2011. 'Tree Species Selection for Land Rehabilitation in Ethiopia: From Fragmented Knowledge to an Integrated Multi-criteria Decision Approach'. *Agroforestry Systems* 82 (3): 303–30. doi:10.1007/s10457-011-9381-8.

Riccioli, Francesco, Roberto Fratini, Fabio Boncinelli, Toufic El Asmar, Jean Pierre El Asmar and Leonardo Casini. 2016. 'Spatial Analysis of Selected Biodiversity Features in Protected Areas: A Case Study in Tuscany Region'. *Land Use Policy* 57. Elsevier Ltd: 540–54. doi:10.1016/j.landusepol.2016.06.023.

Richey, J. S., B. W. Mar and R. R. Horner. 1985. 'Delphi Technique in Environmental Assessment. I. Implementation and Effectiveness'. *Journal of Environmental Management* 21 (2): 135–46.

Ridgley, Mark A, and Gerrit W. Heil. 1998. 'Multicriteria Analysis for Land-Use Management'. 9 (February 2015). doi:10.1007/978-94-015-9058-7. In Euro Beinat and Peter Nijkamp (eds), *Multicriteria Evaluation in Land-Use Management: Methodologies and Case Studies*, vol. 9. Kluwer, 1998. doi:10.1007/978-94-015-9058-7.

Ries, Johannes B. 1996. 'Landscape Damage by Skiing at the Schauinsland in the Black Forest, Germany'. *Mountain Research and Development* 16 (1): 27–40.

Riitters, K. H., R. V. O'Neill and K. B. Jones. 1997. 'Assessing Habitat Suitability at Multiple Scales: A Landscape-Level Approach'. *Biological Conservation* 81 (1–2): 191–202. doi:10.1016/S0006-3207(96)00145-0.

Rixen, C., V. Stoeckli, C. Huovinen and K. Huovinen. 2001. 'The Phenology of Four Subalpine Herbs in Relation to Snow Cover Characteristics'. In A. J. Dolman, A. J. Hall, M. L. Kavvas, T. Oki and J. W. Pomeroy (eds), *Proceedings of the Sixth IAHS Symposium 2001: Soil-Vegetation-Atmosphere Transfer Schemes and Large Scale Hydrological Models*, 359–62. Maastricht: IAHS Publications.

Roy, Bernard. 1996. *Multicriteria Methodology for Decision Aiding*. Boston: MA: Springer US.

Ruth-Balaganskaya, Ekaterina, and Kirsi Myllynen-Malinen. 2000. 'Soil Nutrient Status and Revegetation Practices of Downhill Skiing Areas in Finnish Lapland – A Case Study of Mt. Yllas'. *Landscape and Urban Planning* 50 (4): 259–68. doi:10.1016/S0169-2046(00)00067-0.

Saaty, Thomas L. 1980. *The Analytic Hierarchy Process: Planning, Priority Setting, Resource Allocation*. New York: McGraw-Hill.

Saaty, Thomas L., and Luis G. Vargas, 2012. *Models, Methods, Concepts & Applications of the Analytic Hierarchy Process*. 2nd ed. New York: Springer.

Saltelli, A., F. Campolongo and S. Tarantola. 2000a. 'Sensitivity Analysis as an Ingredient of Modeling'. *Statistical Science* 15 (4): 377–95. doi:10.1214/ss/1009213004.

Saltelli, Andrea, K. Chan and E. M. Scott. 2000b. *Sensitivity Analysis: Wiley Series in Probability and Statistics*. 1st ed. New York: John Wiley & Sons.

Sanecki, Glenn M., Ken Green, Helen Wood and David Lindenmayer. 2006. 'The Implications of Snow-Based Recreation for Small Mammals in the Subnivean Space in South-East Australia'. *Biological Conservation* 129 (4): 511–18. doi:10.1016/j.biocon.2005.11.018.

Schwenk, W. Scott, Therese M. Donovan, William S. Keeton and Jared S. Nunery. 2012. 'Carbon Storage, Timber Production, and Biodiversity: Comparing Ecosystem Services with Multi-criteria Decision Analysis'. *Ecological Applications* 22 (5): 1612–27.

Şener, Başak, M. Lütfi Süzen and Vedat Doyuran. 2006. 'Landfill Site Selection by Using Geographic Information Systems'. *Environmental Geology* 49 (3): 376–88. doi:10.1007/s00254-005-0075-2.

Sharifi, M. A., M. van Herwijnen and W. H. van den Toorn. 2007. 'Spatial Decision Support Systems'. Enschede, The Netherlands: International Institute for Aerospace Survey and Earth Sciences (ITC).

Steinemann, Anne. 2001. 'Improving Alternatives for Environmental Impact Assessment'. *Environmental Impact Assessment Review* 21 (1): 3–21. doi:10.1016/S0195-9255(00)00075-5.

Stöckli, Veronika, and Christian Rixen. 2000. 'Characteristics of Artificial Snow and Its Effect on Vegetation'. Institute for Snow and Avalanche Research SLF, Davos, Switzerland, 468–71.

Strager, Michael P., and Randall S. Rosenberger. 2006. 'Incorporating Stakeholder Preferences for Land Conservation: Weights and Measures in Spatial MCA'. *Ecological Economics* 58 (1): 79–92. doi:10.1016/j.ecolecon.2005.05.024.

Suding, Katharine N., Katherine L. Gross and Gregory R. Houseman. 2004. 'Alternative States and Positive Feedbacks in Restoration Ecology'. *Trends in Ecology and Evolution* 19 (1): 46–53. doi:10.1016/j.tree.2003.10.005.

Thorne, Karen M., Brady J. Mattsson, John Takekawa, Jonathan Cummings, Debby Crouse, Giselle Block, Valary Bloom, et al. 2015. 'Collaborative Decision-Analytic Framework to Maximize Resilience of Tidal Marshes to Climate Change'. *Ecology and Society* 20 (1). doi:10.5751/ES-07018-200130.

Tiitu, Maija, Arto Viinikka, Leena Kopperoinen and Davide Geneletti. 2018. 'Balancing Urban Green Space and Residential Infill Development: A Spatial Multi-Criteria Approach Based on Practitioner Engagement'. *Journal of Environmental Assessment Policy and Management* 20 (30): 1–26.

Tsuyuzaki, Shiro. 2002. 'Vegetation Development Patterns on Skislopes in Lowland Hokkaido, Northern Japan'. *Biological Conservation* 108 (2): 239–46. doi:10.1016/S0006-3207(02)00110-6.

Tversky, Amos, and Daniel Kahneman. 1981. 'The Framing of Decisions and the Psychology of Choice'. *Science* 211 (4481): 453–58.

Urbanska, Krystyna M. 1997. 'Restoration Ecology Research above the Timberline: Colonization of Safety Islands on a Machine-Graded Alpine Ski Run'. *Biodiversity & Conservation* 6 (12): 1655–70.

Uribe, David, Davide Geneletti, Rafael F. del Castillo and Francesco Orsi. 2014. 'Integrating Stakeholder Preferences and GIS-Based Multicriteria Analysis to Identify Forest Landscape Restoration Priorities'. *Sustainability (Switzerland)* 6 (2): 935–51. doi:10.3390/su6020935.

Vallauri, Daniel, James Aronson and Nigel Dudley. 2005. 'An Attempt to Develop a Framework for Restoration Planning'. In S. Mansourian, D. Vallauri and N. Dudley (eds), in cooperation with WWF International, *Forest Restoration in Landscapes: Beyond Planting Trees*, 65–70. New York: Springer.

Vega, Isoida Luna, Othón Alcántara Ayala, David Espinosa Organista and Juan J. Morrone. 1999. 'Historical Relationships of the Mexican Cloud Forests: A Preliminary Vicariance Model Applying Parsimony Analysis of Endemicity to Vascular Plant Taxa'. *Journal of Biogeography* 26 (6): 1299–305. doi:10.1046/j.1365-2699.1999.00361.x.

Villa, Ferdinando, Leonardo Tunesi and Tundi Agardy. 2002. 'Zoning Marine Protected Areas through Spatial Multiple-Criteria Analysis: The Case of the Asinara Island National Marine Reserve of Italy'. *Conservation Biology* 16 (2): 515–26. doi:10.1046/j.1523-1739.2002.00425.x.

von Winterfeldt, D., and W. Edwards. 1986. *Decision Analysis and Behavioral Research*. Cambridge: Cambridge University Press.

Voogd, H. 1983. 'Multicriteria Evaluation for Urban and Regional Planning'. *Pion*, 125. doi:10.6100/IR102252.

Watson, Adam. 1985. 'Soil Erosion and Vegetation Damage near Ski Lifts at Cairn Gorm, Scotland'. *Biological Conservation* 33 (4): 363–81. doi:10.1016/0006-3207(85)90077-1.

Watson, Adam, and Robert Moss. 2004. 'Impacts of Ski-Development on Ptarmigan (Lagopus Mutus) at Cairn Gorm, Scotland'. *Biological Conservation* 116 (2): 267–75. doi:10.1016/S0006-3207(03)00197-6.

Weber, Martin, Franz Eisenführ and Detlof von Winterfeldt. 1988. 'The Effects of Splitting Attributes on Weights in Multiattribute Utility Measurement'. *Management Science* 34 (4): 431–45.

Wichelns, Dennis, James J. Opaluch, Stephen K. Swallow, Thomas F. Weaver and Christopher W. Wessells. 1993. 'A Landfill Site Evaluation Model That Includes Public Preferences Regarding Natural Resources and Nearby Communities'. *Waste Management & Research* 11 (3).

Wipf, Sonja, Christian Rixen, Markus Fischer, Bernhard Schmid and Veronika Stoeckli. 2005. 'Effects of Ski Piste Preparation on Alpine Vegetation'. *Journal of Applied Ecology* 42 (2): 306–16. doi:10.1111/j.1365-2664.2005.01011.x.

World Conservation Monitoring Centre (WCMC). 2000. 'Prioritisation of Target Areas for Forest Restoration: Final Report'. Cambridge: World Conservation Monitoring Centre.

Yoon, Paul K., and Ching-Lai Hwang. 1995. *Multiple Attribute Decision Making: An Introduction.* London: SAGE Publications.

Zhang, Zhiming, Ruth Sherman, Zijiang Yang, Ruidong Wu, Wenli Wang, Mei Yin, Guihua Yang and Xiaokun Ou. 2013. 'Integrating a Participatory Process with a GIS-Based Multi-criteria Decision Analysis for Protected Area Zoning in China'. *Journal for Nature Conservation* 21 (4). Elsevier GmbH: 225–40. doi:10.1016/j.jnc.2012.12.006.

Zia, Asim, Paul Hirsch, Alexander Songorwa, David R. Mutekanga, Sheila O'Connor, Thomas McShane, Peter Brosius and Bryan Norton. 2011. 'Cross-Scale Value Trade-Offs in Managing Social-Ecological Systems: The Politics of Scale in Ruaha National Park, Tanzania'. *Ecology and Society* 16 (4). doi:10.5751/ES-04375-160407.

INDEX

www.ingramcontent.com/pod-product-compliance
Ingram Content Group UK Ltd.
Pitfield, Milton Keynes, MK11 3LW, UK
UKHW041832100926
8556IPUK00002B/126

9 781783 089529